T0213947

Lecture Notes in Computer Science   12094

More information about this series at http://www.springer.com/series/7407

Daniela Petrişan · Jurriaan Rot (Eds.)

Coalgebraic Methods in Computer Science

15th IFIP WG 1.3 International Workshop, CMCS 2020
Colocated with ETAPS 2020
Dublin, Ireland, April 25–26, 2020
Proceedings

 Springer

Editors
Daniela Petrişan
Université de Paris, CNRS, IRIF
Paris, France

Jurriaan Rot
Radboud University
Nijmegen, The Netherlands

ISSN 0302-9743   ISSN 1611-3349   (electronic)
Lecture Notes in Computer Science
ISBN 978-3-030-57200-6   ISBN 978-3-030-57201-3   (eBook)
https://doi.org/10.1007/978-3-030-57201-3

LNCS Sublibrary: SL1 – Theoretical Computer Science and General Issues

This Springer imprint is published by the registered company Springer Nature Switzerland AG
The registered company address is: Gewerbestrasse 11, 6330 Cham, Switzerland

Preface

The 15th International Workshop on Coalgebraic Methods in Computer Science (CMCS 2020) was originally planned to be held on April 25–26, 2020, in Dublin, Ireland, as a satellite event of the European Joint Conferences on Theory and Practice of Software (ETAPS 2020). Due to the 2020 COVID-19 pandemic, it was canceled, and held in alternative form as a series of online seminars, ranging over several weeks in the fall of 2020.

The aim of the workshop is to bring together researchers with a common interest in the theory of coalgebras, their logics, and their applications. Coalgebras allow for a uniform treatment of a large variety of state-based dynamical systems, such as transition systems, automata (including weighted and probabilistic variants), Markov chains, and game-based systems. Over the last two decades, coalgebra has developed into a field of its own interest, presenting a deep mathematical foundation, a growing field of applications, and interactions with various other fields such as reactive and interactive system theory, object-oriented and concurrent programming, formal system specification, modal and description logics, artificial intelligence, dynamical systems, control systems, category theory, algebra, analysis, etc.

Previous workshops of the CMCS series were held in Lisbon (1998), Amsterdam (1999), Berlin (2000), Genova (2001), Grenoble (2002), Warsaw (2003), Barcelona (2004), Vienna (2006), Budapest (2008), Paphos (2010), Tallinn (2012), Grenoble (2014), Eindhoven (2016), and Thessaloniki (2018). Since 2004, CMCS has been a biennial workshop, alternating with the International Conference on Algebra and Coalgebra in Computer Science (CALCO).

The CMCS 2020 program featured a keynote talk by Yde Venema (ILLC, University of Amsterdam, The Netherlands), an invited talk by Nathanaël Fijalkow (CNRS, LaBRI, University of Bordeaux, France), and an invited talk by Koko Muroya (RIMS, Kyoto University, Japan). In addition, a special session on probabilistic couplings was planned, featuring invited tutorials by Marco Gaboardi (Boston University, USA) and Justin Hsu (University of Wisconsin-Madison, USA). This volume contains revised regular contributions (9 papers were accepted out of 13 submissions; 1 submission was conditionally accepted but unfortunately had to be rejected ultimately as the author did not agree to implement the changes required by the Program Committee) and the abstracts of 3 keynote/invited talks. All regular contributions were refereed by three reviewers. We wish to thank all the authors who submitted to CMCS 2020, and the external reviewers and the Program Committee members for their thorough reviewing and help in improving the papers accepted at CMCS 2020.

July 2020

Daniela Petrişan
Jurriaan Rot

Organization

Program Committee Chairs

Daniela Petrişan   Université de Paris, CNRS, IRIF, France
Jurriaan Rot   Radboud University, The Netherlands

Steering Committee

Filippo Bonchi   University of Pisa, Italy
Marcello Bonsangue   CWI/Leiden University, The Netherlands
Corina Cîrstea   University of Southampton, UK
Ichiro Hasuo   National Institute of Informatics, Japan
Bart Jacobs   Radboud University, The Netherlands
Bartek Klin   University of Warsaw, Poland
Alexander Kurz   Chapman University, USA
Marina Lenisa   University of Udine, Italy
Stefan Milius (Chair)   FAU Erlangen-Nürnberg, Germany
Larry Moss   Indiana University, USA
Dirk Pattinson   The Australian National University, Australia
Lutz Schröder   FAU Erlangen-Nürnberg, Germany
Alexandra Silva   University College London, UK

Program Committee

Henning Basold   Leiden University, The Netherlands
Nick Bezhanishvili   University of Amsterdam, The Netherlands
Corina Cîrstea   University of Southampton, UK
Mai Gehrke   CNRS and Université Côte d'Azur, France
Helle Hvid Hansen   Delft University of Technology, The Netherlands
Shin-Ya Katsumata   National Institute of Informatics, Japan
Bartek Klin   Warsaw University, Poland
Ekaterina Komendantskaya   Heriot-Watt University, UK
Barbara König   Universität Duisburg-Essen, Germany
Dexter Kozen   Cornell University, USA
Clemens Kupke   University of Strathclyde, UK
Alexander Kurz   Chapman University, USA
Daniela Petrişan   Université de Paris, France
Andrei Popescu   Middlesex University London, UK
Damien Pous   CNRS, ENS Lyon, France
Jurriaan Rot   Radboud University, The Netherlands
Davide Sangiorgi   University of Bologna, Italy

Ana Sokolova	University of Salzburg, Austria
David Sprunger	National Institute of Informatics, Japan
Henning Urbat	FAU Erlangen-Nürnberg, Germany
Fabio Zanasi	University College London, UK

Additional Reviewers

Alexandre Goy
Gerco van Heerdt
Stefan Milius
Laureline Pinault
Luigi Santocanale

Sponsoring Institutions

IFIP WG 1.3
Logical Methods in Computer Science e.V.

Abstracts of Invited Talks

Logic and Automata: A Coalgebraic Perspective

Yde Venema

Institute for Logic, Language and Computation, Universiteit van Amsterdam
y.venema@uva.nl
https://staff.fnwi.uva.nl/y.venema/

Abstract. In Theoretical Computer Science, the area of *Logic and Automata* combines a rich mathematical theory with many applications in for instance the specification and verification of software. Of particular relevance in this area is the design of formal languages and derivation systems for describing and reasoning about nonterminating or *ongoing behavior*, and the study of such formalisms through the theory of automata operating on potentially infinite objects.

We will take a foundational approach towards these issues using insights from Universal Coalgebra and Coalgebraic Logic. Specifically, we will review the basic theory of *coalgebraic modal fixpoint logic* and its link with *coalgebra automata* – finite automata that operate on coalgebras.We will show that much of the theory linking logic and automata in the setting of streams, trees or transition systems, transfers to the coalgebraic level of generality, and we will argue that this theory is essentially coalgebraic in nature.

Topics that will be covered include closure properties of classes of coalgebra automata, expressive completeness results concerning fixpoint logics, and the design of sound and complete derivation systems for modal fixpoint logics. The key instruments in our analysis will be those of a coalgebraic modal signature and its associated one-step logic. As we will see, many results in the theory of logic and automata are ultimately grounded in fairly simple observations on the underlying one-step logic.

Keywords: Coalgebra · Modal logic · Fixed points · Automata

The Theory of Universal Graphs for Games: Past and Future

Nathanaël Fijalkow[1,2] (iD)

[1] CNRS, LaBRI, Université de Bordeaux, Bordeaux, France
nathanael.fijalkow@labri.fr
[2] The Alan Turing Institute of Data Science, London, UK

Abstract. This paper surveys recent works about the notion of universal graphs. They were introduced in the context of parity games for understanding the recent quasipolynomial time algorithms, but they are defined for arbitrary objectives yielding a new approach for constructing efficient algorithms for solving different classes of games.

Keywords: Games · Parity games · Universal graphs

Hypernet Semantics and Robust Observational Equivalence

Koko Muroya

RIMS, Kyoto University, Kyoto, Japan
kmuroya@kurims.kyoto-u.ac.jp

Abstract. In semantics of programming languages, *observational equivalence* [6] is the fundamental and classical question, asking whether two program fragments have the same behaviour in any program context. Establishing observational equivalence is vital in validating compiler optimisation and refactoring, and there have been proposed proof methodologies for observational equivalence, such as logical relations [9, 10] and applicative bisimulations [1]. A challenge is, however, to establish *robustness* of observational equivalence, namely to prove that observational equivalence is preserved by a language extension. In this talk I will discuss how *hypernet semantics* can be used to reason about such robustness. The semantics was originally developed for cost analysis of program execution [8], in particular for analysis of a trade-off between space and time efficiency. It has been used to model exotic programming features [2, 7] inspired by Google's TensorFlow[1], a machine learning library, and also to design a dataflow programming language [3]. In hypernet semantics, program execution is modelled as dynamic rewriting of a *hypernet*, that is, graph representation of a program. Inspired by Girard's Geometry of Interaction [4], the rewriting process is notably guided and controlled by a dedicated object (*token*) of the graph. The use of graphs together with the token enables a direct proof of observational equivalence. The key step of the proof is to establish *robustness* of observational equivalence relative to language features, using step-wise reasoning enriched with the so-called *up-to* technique [5].

References

1. Abramsky, S.: The Lazy Lambda-Calculus, pp. 65–117. Addison Wesley (1990)
2. Cheung, S., Darvariu, V., Ghica, D.R., Muroya, K., Rowe, R.N.S.: A functional perspective on machine learning via programmable induction and abduction. In: Gallagher, J., Sulzmann, M. (eds.) FLOPS 2018. LNCS, vol. 10818. Springer, Cham (2018). https://doi.org/10.1007/978-3-319-90686-7_6
3. Cheung, S.W.T., Ghica, D.R., Muroya, K.: Transparent synchronous dataflow. arXiv preprint arXiv:1910.09579 (2019)
4. Girard, J.Y.: Geometry of interaction I: interpretation of system F. In: Logic Colloquium 1988. Studies in Logic and the Foundations of Mathematics, vol. 127, pp. 221–260. Elsevier (1989). https://doi.org/10.1016/S0049-237X(08)70271-4

---

[1] https://www.tensorow.org/.

5. Milner, R.: Communicating and Mobile Systems: the Pi-calculus. Cambridge University Press (1999)
6. Morris, Jr., J.H.: Lambda-calculus models of programming languages. Ph.D. thesis, Massachusetts Institute of Technology (1969)
7. Muroya, K., Cheung, S.W.T., Ghica, D.R.: The geometry of computation-graph abstraction. In: LICS 2018, pp. 749–758. ACM (2018). https://doi.org/10.1145/3209108.3209127
8. Muroya, K., Ghica, D.R.: The dynamic geometry of interaction machine: a token-guided graph rewriter. LMCS 15(4) (2019). https://lmcs.episciences.org/5882
9. Plotkin, G.D.: Lambda-Definability and Logical Relations. Memorandum SAI-RM-4 (1973)
10. Statman, R.: Logical relations and the typed lambda-calculus. Inf. Control 65(2/3), 85–97 (1985). https://doi.org/10.1016/S0019-9958(85)80001-2

Contents

The Theory of Universal Graphs for Games: Past and Future

Nathanaël Fijalkow[1,2]([⊠])

[1] CNRS, LaBRI, Université de Bordeaux, Bordeaux, France
nathanael.fijalkow@labri.fr
[2] The Alan Turing Institute of Data Science, London, UK

Abstract. This paper surveys recent works about the notion of universal graphs. They were introduced in the context of parity games for understanding the recent quasipolynomial time algorithms, but they are defined for arbitrary objectives yielding a new approach for constructing efficient algorithms for solving different classes of games.

Keywords: Games · Parity games · Universal graphs

1 The Quasipolynomial Era for Parity Games

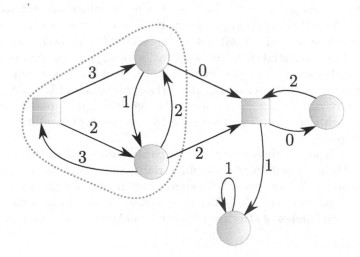

Fig. 1. A parity game. The vertices controlled by Eve are represented by circles and the ones controlled by Adam as squares. The edges are labelled by priorities inducing the parity objective. The dotted region is the winning region for Eve.

Parity games are a central model in the study of logic and automata over infinite trees and for their tight relationship with model-checking games for the modal

© IFIP International Federation for Information Processing 2020
Published by Springer Nature Switzerland AG 2020
D. Petrişan and J. Rot (Eds.): CMCS 2020, LNCS 12094, pp. 1–10, 2020.
https://doi.org/10.1007/978-3-030-57201-3_1

μ-calculus. The central question left open for decades is whether there exists a polynomial time algorithm for solving parity games. This section is a historical account on the progress made on this question since 2017 (Fig. 1).

The breakthrough result of Calude, Jain, Khoussainov, Li, and Stephan [2] was to construct a quasipolynomial time algorithm for solving parity games. Following decades of exponential and subexponential algorithms, this very surprising result triggered further research: soon after further quasipolynomial time algorithms were constructed reporting almost the same complexity, which is roughly $n^{O(\log d)}$. Let us classify them in two families.

The first family of algorithms includes the original algorithm by Calude, Jain, Khoussainov, Li, and Stephan [2] (see also [6] for a presentation of the algorithm as value iteration), then the succinct progress measure algorithm by Jurdziński and Lazić [9] (see also [7] for a presentation of the algorithm using universal trees explicitly) and the register games algorithm by Lehtinen [11] (see also [14] for a presentation of the algorithm using good-for-small-games automata explicitly). Bojańczyk and Czerwiński [1] introduced the separation question, describing a family of algorithms for solving parity games based on *separating automata*, and showed that the first quasipolynomial time algorithm yields a quasipolynomial solution to the separation question. Later Czerwiński, Daviaud, Fijalkow, Jurdziński, Lazić, and Parys [5] showed that the other two algorithms also yield quasipolynomial solutions to the separation question. The main contribution of [5] is to show that any separating automaton contains a *universal tree* in its set of states; in other words, the three algorithms in this first family induce three (different) constructions of universal trees.

The second family of algorithms is so-called Zielonka-type algorithms, inspired by the exponential time algorithm by Zielonka [15]. The first quasipolynomial time algorithm is due to Parys [13], its complexity was improved by Lehtinen, Schewe, and Wojtczak [12]. Recently Jurdziński and Morvan [10] constructed a universal attractor decomposition algorithm encompassing all three algorithms: each algorithm is parametrised by the choice of two universal trees (one of each player).

All quasipolynomial time algorithms for parity games fall in one of two families, and both are based on the combinatorial notion of universal trees. This notion is by now well understood, with almost matching upper and lower bounds on the size of universal trees [5,7]. The lower bound implies a complexity barrier applying to both families of algorithms, hence to all known quasipolynomial time algorithms.

2   Universal Graphs

Colcombet and Fijalkow [3,4] revisited the relationship between separating automata and universal trees by introducing the notions of *good-for-small-games automata* and of *universal graphs*.

The notion of good-for-small-games automata extends separating automata and resolves a disturbing situation that we describe now. The first three quasipolynomial time algorithms discussed above construct separating automata, but for the third algorithm the automaton is *non-deterministic*. For that reason its correctness does not follow from the generic argument applying to the class of algorithms based on *deterministic* separating automata. (We note however that the lower bound result of [5] does apply to non-deterministic separating automata.) The notion of good-for-small-games automata resolves this conundrum, since the automaton constructed in Lehtinen's algorithm is indeed good-for-small-games and the generic argument for deterministic separating automata naturally extends to this class of automata[1].

Universal trees arise in the study of the parity objective, the tree structure representing the nested behaviour of priorities. The notion of *universal graphs* extends universal trees from parity objectives to arbitrary (positionally determined) objectives. The main result of [4] is an equivalence result between *good-for-small-games automata* and *universal graphs*. More specifically, a good-for-small-games automaton induces a universal graph of the same size, and conversely. This equivalence extends the results of [5] relating separating automata and universal trees to any positionally determined objectives, so in particular for parity and mean payoff games.

The remainder of this section gives the formal definitions of universal graphs, good-for-small-games automata, and separating automata, their use for solving games, and state the equivalence result.

Universal Graphs

We deliberately postpone the definitions related to games: indeed the notion of universal graphs is a purely combinatorial notion on graphs and ignores the interactive aspects of games.

Graphs. We consider edge labelled directed graphs: a graph is given by a (finite) set V of vertices and a (finite) set $E \subseteq V \times C \times V$ of edges. A vertex v for which there exists no outgoing edge $(v, \ell, v') \in E$ is called a sink. We let n denote the number of vertices and m the number of edges. The size of a graph is its number of vertices.

Homomorphisms. For two graphs G, G', a homomorphism $\phi : G \to G'$ maps the vertices of G to the vertices of G' such that

$$(v, \ell, v') \in E \implies (\phi(v), \ell, \phi(v')) \in E'.$$

As a simple example that will be useful later on, note that if G' is a super graph of G, meaning they have the same set of vertices and every edge in G is also in G', then the identity is a homomorphism $G \to G'$.

Paths. A path π is a (finite or infinite) sequence of consecutive edges, where consecutive means that the third component of a triple in the sequence matches

---

[1] A closely related solution was given in [14] by giving syntactic properties implying that a non-deterministic automaton is good-for-small-games and showing that the automaton constructed in Lehtinen's algorithm has this property.

the first component of the next triple. In the case of a finite path we write $last(\pi)$ for the last vertex in π. We write $\pi = (v_0, \ell_0, v_1)(v_1, \ell_1, v_2) \cdots$ and let $\pi_{\leq i}$ denote the prefix of π of length i, meaning $\pi_{\leq i} = (v_0, \ell_0, v_1) \cdots (v_{i-1}, \ell_{i-1}, v_i)$.

Objectives. An objective is a set $\Omega \subseteq C^\omega$ of infinite sequences of colours. A sequence of colours belonging to Ω is said to satisfy Ω. We say that a path in a graph satisfies Ω, or that it is winning when Ω is clear from context, if the sequence of labels it visits belongs to Ω. A path is said to be maximal if it is either infinite or ends in a sink.

We will work with a generic objective Ω. However it is convenient to assume that Ω is *prefix independent*. Formally, this means that $\Omega = C^* \cdot \Omega$, or equivalently for all $w \in C^*$ and $\rho \in C^\omega$, we have $\rho \in \Omega \iff w \cdot \rho \in \Omega$. This assumption can be lifted at the price of working with graphs with a distinguished initial vertex v_0 which must be preserved by homomorphisms.

Definition 1 (Graphs satisfying an objective). *Let Ω be a prefix independent objective. A graph satisfies Ω if all maximal paths are infinite and winning.*

Note that if a graph contains a sink, it does not satisfy Ω because it contains some finite maximal path.

Definition 2 (Universal graphs). *Let Ω be a prefix independent objective. A graph $\mathcal{U}$ is (n, Ω)-universal if it satisfies the following two properties:*

1. *it satisfies Ω,*
2. *for any graph G of size at most n satisfying Ω, there exists a homomorphism $\phi : G \to \mathcal{U}$.*

It is not clear that for any objective Ω and $n \in \mathbb{N}$ there exists an (n, Ω)-universal graph. Indeed, the definition creates a tension between "satisfying Ω", which restricts the structure, and "homomorphically mapping any graph of size at most n satisfying Ω", implying that the graph witnesses many different behaviours. A simple construction of an (n, Ω)-universal graph is to take the disjoint union of all (n, Ω)-graphs satisfying Ω. Since up to renaming of vertices there are finitely many such graphs this yields a very large but finite (n, Ω)-universal graph.

When considering a universal graph $\mathcal{U}$ we typically let $V_{\mathcal{U}}$ and $E_{\mathcal{U}}$ denote the sets of vertices and edges of $\mathcal{U}$, respectively.

Solving Games Using Universal Graphs by Reduction to Safety Games

We now introduce infinite duration games on graphs and construct a family of algorithms for solving games using universal graphs by reduction to safety games [1,4].

Arenas. An arena is given by a graph *containing no sink* together with a partition $V_{\text{Eve}} \uplus V_{\text{Adam}}$ of its set V of vertices describing which player controls each vertex. We represent vertices controlled by Eve with circles and those controlled by Adam with squares.

Games. A game is given by an arena and an objective. We often let $\mathcal{G}$ denote a game, its size is the size of the underlying graph. It is played as follows. A token is initially placed on some initial vertex v_0, and the player who controls this vertex pushes the token along an edge, reaching a new vertex; the player who controls this new vertex takes over, and this interaction goes on forever describing an infinite path.

Strategies. We write Path_{v_0} for the set of finite paths of $\mathcal{G}$ starting from v_0 and ending in V_{Eve}. A strategy for Eve is a map $\sigma : \mathrm{Path}_{v_0} \to E$ such that for all $\pi \in \mathrm{Path}_{v_0}$, $\sigma(\pi)$ is an edge in E from $\mathrm{last}(\pi)$. Note that we always take the point of view of Eve, so a strategy implicitly means a strategy of Eve, and winning means winning for Eve. We say that an infinite path $\pi = (v_0, \ell_0, v_1)(v_1, \ell_1, v_2) \cdots$ is consistent with the strategy σ if for all i, if $v_i \in V_{\mathrm{Eve}}$, then $\sigma(\pi_{\leq i}) = (v_i, \ell_i, v_{i+1})$.

A strategy σ is winning from v_0 if all infinite paths starting from v_0 and consistent with σ are winning. Solving a game is the following decision problem:

INPUT: a game $\mathcal{G}$ and an initial vertex v_0.
OUTPUT: "yes" if Eve has a winning strategy from v_0, "no" otherwise.

We say that v_0 is a winning vertex of $\mathcal{G}$ when the answer to the above problem is positive. The winning region of Eve of $\mathcal{G}$ is the set of winning vertices.

Positional Strategies. Positional strategies make decisions only considering the current vertex: $\sigma : V_{\mathrm{Eve}} \to E$. A positional strategy induces a strategy $\widehat{\sigma} : \mathrm{Path}_{v_0} \to E$ by $\widehat{\sigma}(\pi) = \sigma(\mathrm{last}(\pi))$, where by convention the last vertex of the empty path is the initial vertex v_0.

Definition 3 (Positionally determined objectives). *We say that an objective Ω is positionally determined if for every game with objective Ω and initial vertex v_0, whenever there exists a winning strategy from v_0 then there exists a positional winning strategy from v_0.*

Given a game $\mathcal{G}$, an initial vertex v_0, and a positional strategy σ we let $\mathcal{G}[\sigma, v_0]$ denote the graph obtained by restricting $\mathcal{G}$ to vertices reached by σ from v_0 and to the moves prescribed by σ. Formally, the set of vertices and edges is

$$V[\sigma, v_0] = \{v \in V : \text{there exists a path from } v_0 \text{ to } v \text{ consistent with } \sigma\},$$
$$E[\sigma, v_0] = \{(v, \ell, v') \in E : v \in V_{\mathrm{Adam}} \text{ or } (v \in V_{\mathrm{Eve}} \text{ and } \sigma(v) = (v, \ell, v'))\}$$
$$\cap \ V[\sigma, v_0] \times C \times V[\sigma, v_0].$$

Fact 1. *Let Ω be a prefix independent objective, $\mathcal{G}$ be a game, v_0 be an initial vertex, and σ a positional strategy. Then the strategy σ is winning from v_0 if and only if the graph $\mathcal{G}[\sigma, v_0]$ satisfies Ω.*

Safety Games. The safety objective `Safe` on two colours $C = \{\varepsilon, \mathrm{Lose}\}$ is given by $\mathtt{Safe} = \{\varepsilon^\omega\}$. In words, an infinite path is winning if it avoids the letter Lose. The following lemma is folklore.

Lemma 1. *Given a safety game with n vertices and m edges, there exists an algorithm running in time and space $O(n + m)$ which computes the winning region of Eve.*

Consider a game $\mathcal{G}$ with objective Ω of size n and an (n, Ω)-universal graph $\mathcal{U}$, we construct a safety game $\mathcal{G} \rhd \mathcal{U}$ as follows. The arena for the game $\mathcal{G} \rhd \mathcal{U}$ is given by the following set of vertices and edges

$$
\begin{aligned}
V'_{\text{Eve}} &= V_{\text{Eve}} \times V_{\mathcal{U}} \uplus V \times V_{\mathcal{U}} \times C, \\
V'_{\text{Adam}} &= V_{\text{Adam}} \times V_{\mathcal{U}}, \\
E' &= \{((v, s), \varepsilon, (v', s, \ell)) : (v, \ell, v') \in E\} \\
&\cup \{((v', s, \ell), \varepsilon, (v', s')) : (s, \ell, s') \in E_{\mathcal{U}}\} \\
&\cup \{((v, s, \ell), \text{Lose}, (v, s, \ell)) : v \in V_{\mathcal{G}}, s \in V_{\mathcal{U}}, \ell \in C\}.
\end{aligned}
$$

In Words: the game $\mathcal{G} \rhd \mathcal{U}$ simulates $\mathcal{G}$. From (v, s), the following two steps occur. First, the player who controls v picks an edge $(v, \ell, v') \in E$ as he would in $\mathcal{G}$ which leads to (v', s, ℓ), and second, Eve chooses which edge (s, ℓ, s') to follow in the universal graph $\mathcal{U}$ which leads to (v', s'). If Eve is unable to play in $\mathcal{U}$ (because there are no outgoing edges of the form (s, ℓ, s')), she is forced to choose $((v, s, \ell), \text{Lose}, (v, s, \ell))$ and lose the safety game.

Theorem 1. *Let Ω be a prefix independent positionally determined objective. Let $\mathcal{G}$ be a game of size n with objective Ω and $\mathcal{U}$ an (n, Ω)-universal graph. Let v_0 be an initial vertex. Then Eve has a winning strategy in $\mathcal{G}$ from v_0 if and only if there exists a vertex s_0 in $\mathcal{U}$ such that she has a winning strategy in $\mathcal{G} \rhd \mathcal{U}$ from (v_0, s_0).*

This theorem can be used to reduce games with objective Ω to safety games, yielding an algorithm whose complexity is proportional to the number of edges of $\mathcal{U}$. Indeed, recall that the complexity of solving a safety game is proportional to the number of vertices plus edges. The number of vertices of $\mathcal{G} \rhd \mathcal{U}$ is $O(n \cdot n_{\mathcal{U}}|C|)$, and the number of edges is $O(n \cdot m_{\mathcal{U}})$ so the overall complexity is $O(n \cdot (n_{\mathcal{U}}|C| + m_{\mathcal{U}}))$.

Proof. Let us assume that Eve has a winning strategy σ in $\mathcal{G}$ from v_0, which can be chosen to be positional, Ω being positionally determined. We consider the graph $\mathcal{G}[\sigma, v_0]$: it has size at most n and since σ is winning from v_0 it satisfies Ω. Because $\mathcal{U}$ is an (n, Ω)-universal graph there exists a homomorphism ϕ from $\mathcal{G}[\sigma, v_0]$ to $\mathcal{U}$. We construct a winning strategy in $\mathcal{G} \rhd \mathcal{U}$ from $(v_0, \phi(v_0))$ by playing as in σ in the first component and following the homomorphism ϕ on the second component.

Conversely, let σ be a winning strategy from (v_0, s_0) in $\mathcal{G} \rhd \mathcal{U}$. Consider the strategy σ' in $\mathcal{G}$ which mimics σ by keeping an additional component in $\mathcal{U}$ which is initially set to s_0 and updated following σ. The strategy σ' is winning since all infinite paths in $\mathcal{U}$ satisfy Ω.

Good-for-Small-Games Automata

The automata we consider are non-deterministic safety automata over infinite words on the alphabet C, where safety means that all states are accepting: a word is rejected if there exist no run for it. Formally, an automaton is given by a set of states Q, an initial state $q_0 \in Q$, and a transition relation $\Delta \subseteq Q \times C \times Q$. A run is a finite or infinite sequence of consecutive transitions starting with the initial state q_0, where consecutive means that the third component of a triple in the sequence matches the first component of the next triple. We say that it is a run over a word if the projection of the run on the colours matches the word. An infinite word is accepted if it admits an infinite run. We let $L(\mathcal{A})$ denote the set of infinite words accepted by an automaton $\mathcal{A}$.

We introduce the notion of a *guiding strategy* for resolving non-determinism. A guiding strategy is a function $\sigma : C^+ \to \Delta$: given a finite word $\rho \in C^+$ it picks a transition. More specifically for an infinite word $\rho = c_0 c_1 \cdots \in C^\omega$ the guiding strategy chooses the following infinite sequence of transitions:

$$\sigma(c_0)\sigma(c_0 c_1)\sigma(c_0 c_1 c_2)\dots.$$

It may fail to be a run over ρ even if there exists one: the choice of transitions may require knowing the whole word ρ in advance, and the guiding strategy has to pick the transition only based on the current prefix. We say that the automaton guided by σ accepts a word ρ if the guiding strategy indeed induces an infinite run.

In the following we say that a path in a graph is accepted or rejected by an automaton; this is an abuse of language since what the automaton reads is only the sequence of colours of the corresponding path.

Definition 4. *An automaton is (n, Ω)-good-for-small-games if the two following properties hold:*

1. *there exists a guiding strategy σ such that for all graphs G of size n satisfying Ω, the automaton guided by σ accepts all paths in the graph G,*
2. *the automaton rejects all paths not satisfying Ω.*

In a very similar way as universal graphs, good-for-small-games automata can be used to construct reductions to safety games.

Consider a game $\mathcal{G}$ with objective Ω of size n and an (n, Ω)-good-for-small-games $\mathcal{A}$, we construct a safety game $\mathcal{G} \triangleright \mathcal{A}$ as follows. We let Q denote the set of states of $\mathcal{A}$ and Δ the transition relation. The arena for the game $\mathcal{G} \triangleright \mathcal{A}$ is given by the following set of vertices and edges

$$
\begin{aligned}
V'_{\text{Eve}} &= V_{\text{Eve}} \times Q \uplus V_{\text{Eve}} \times Q \times C, \\
V'_{\text{Adam}} &= V_{\text{Adam}} \times Q, \\
E' &= \{((v,q), \varepsilon, (v',q,\ell) : (v,\ell,v') \in E\} \\
&\quad \cup \{((v',q,\ell), \varepsilon, (v',q')) : (q,\ell,q') \in \Delta\} \\
&\quad \cup \{((v,q,\ell), \text{Lose}, (v,q,\ell)) : v \in V, q \in Q, \ell \in C\}.
\end{aligned}
$$

In words: the game $\mathcal{G} \triangleright \mathcal{A}$ simulates $\mathcal{G}$. From (v,q), the following two steps occur. First, the player who controls v picks an edge $(v,\ell,v') \in E$ as he would in $\mathcal{G}$ which leads to (v',q,ℓ), and second, Eve chooses which transition (q,ℓ,q') to follow in the automaton $\mathcal{A}$ which leads to (v',q'). If Eve is unable to choose a transition in $\mathcal{A}$ (because there are no transitions of the form (q,ℓ,q')), she is forced to choose $((v,q,\ell), \text{Lose}, (v,q,\ell))$ and lose the safety game.

Theorem 2. *Let Ω be a prefix independent positionally determined objective. Let $\mathcal{G}$ be a game of size n with objective Ω and $\mathcal{A}$ an (n, Ω)-good-for-small-games automaton. Let v_0 be an initial vertex. Then Eve has a winning strategy in $\mathcal{G}$ from v_0 if and only if she has a winning strategy in $\mathcal{G} \triangleright \mathcal{A}$ from (v_0, q_0).*

As for universal graphs, this theorem can be used to reduce games with objective Ω to safety games, and the complexity is dominated by the number of transitions of $\mathcal{A}$.

Separating Automata

Definition 5. *An (n, Ω)-separating automaton is a deterministic (n, Ω)-good-for-small-games automaton.*

In the deterministic case the first property reads: for all graphs G of size n satisfying Ω, the automaton accepts all paths in the graph G. The advantage of using deterministic automata is that the resulting safety game is smaller, and indeed the complexity of the algorithm above is proportional to the number of states of $\mathcal{A}$, more specifically it is $O(n \cdot |Q|)$ where Q is the set of states of $\mathcal{A}$.

The Equivalence Result

Theorem 3 ([4]). *Let Ω be a prefix independent positionally determined objective and n a natural number.*

- *For any (n, Ω)-separating automaton one can construct an (n, Ω)-good-for-small-games automaton of the same size.*
- *For any (n, Ω)-good-for-small-games automaton one can construct an (n, Ω)-universal graph of the same size.*
- *For any (n, Ω)-universal graph one can construct an (n, Ω)-separating automaton of the same size.*

The first item is trivial because by definition a separating automaton is a good-for-small-games automaton. The other two constructions are non-trivial, and in particular it is not true that a good-for-small-games automaton directly yields a universal graph as is, and the same holds conversely. Both constructions are based on a saturation method which constructs a linear order either on the set of states or on the set of vertices.

As a corollary of this equivalence result we can refine the reduction to safety games and present it in a value iteration style which has the same time complexity but also achieves a very small space complexity. We refer to [8] for more details.

3   Perspectives

The notion of universal graphs is a new tool for constructing algorithms for several classes of games by reducing them to safety games. For each objective, the question becomes whether there exist small universal graphs. A benefit of this approach is that this becomes a purely combinatorial question about graphs which does not involve reasoning about games.

Let us briefly discuss some of the results obtained so far, and gives some perspectives for future research. The first example is parity objectives. In that case universal graphs have been combinatorially well understood: they are equivalently described as universal trees and their size is quasipolynomial with almost matching upper and lower bounds [7]. A more systematic study of ω-regular objectives reveals that similar ideas lead to small universal graphs for objectives such as Rabin or disjunctions of parity objectives. The second example is mean payoff objectives, which extend parity objectives. Upper and lower bounds have been obtained recently for this class of objectives, yielding new algorithms with (slightly) improved complexity [8].

In all examples mentioned above, the algorithms obtained using the universal graph technology match the best known complexity for solving these games.

Universal Graphs for Families of Graphs
One can easily extend the theory of universal graphs to consider games over a given family of graphs. This motivates the question whether there exist small universal graphs for different families of graphs such as graphs of bounded tree or clique width.

Combining Objectives: Compositional Constructions of Universal Graphs
An appealing question is whether universal graphs can be constructed compositionally. Indeed, it is often useful to consider objectives which are described as boolean combinations of objectives. Can we give generic constructions using universal graphs for basic objectives to build universal graphs for more complicated objectives? As examples, let us mention unions of mean payoff objectives (with both infimum limit or supremum limit semantics) and disjunctions of parity and mean payoff objectives.

References

1. Bojańczyk, M., Czerwiński, W.: An automata toolbox, February 2018. https://www.mimuw.edu.pl/~bojan/papers/toolbox-reduced-feb6.pdf
2. Calude, C.S., Jain, S., Khoussainov, B., Li, W., Stephan, F.: Deciding parity games in quasipolynomial time. In: STOC, pp. 252–263 (2017). https://doi.org/10.1145/3055399.3055409
3. Colcombet, T., Fijalkow, N.: Parity games and universal graphs. CoRR abs/1810.05106 (2018)

4. Colcombet, T., Fijalkow, N.: Universal graphs and good for games automata: new tools for infinite duration games. In: FoSSaCS, pp. 1–26 (2019). https://doi.org/10.1007/978-3-030-17127-8_1

5. Czerwiński, W., Daviaud, L., Fijalkow, N., Jurdziński, M., Lazić, R., Parys, P.: Universal trees grow inside separating automata: quasi-polynomial lower bounds for parity games. CoRR abs/1807.10546 (2018)

6. Fearnley, J., Jain, S., Schewe, S., Stephan, F., Wojtczak, D.: An ordered approach to solving parity games in quasi polynomial time and quasi linear space. In: SPIN, pp. 112–121 (2017)

7. Fijalkow, N.: An optimal value iteration algorithm for parity games. CoRR abs/1801.09618 (2018)

8. Fijalkow, N., Gawrychowski, P., Ohlmann, P.: The complexity of mean payoff games using universal graphs. CoRR abs/1812.07072 (2018)

9. Jurdziński, M., Lazić, R.: Succinct progress measures for solving parity games. In: LICS, pp. 1–9 (2017)

10. Jurdziński, M., Morvan, R.: A universal attractor decomposition algorithm for parity games. CoRR abs/2001.04333 (2020)

11. Lehtinen, K.: A modal-μ perspective on solving parity games in quasi-polynomial time. In: LICS, pp. 639–648 (2018)

12. Lehtinen, K., Schewe, S., Wojtczak, D.: Improving the complexity of Parys' recursive algorithm. CoRR abs/1904.11810 (2019)

13. Parys, P.: Parity games: Zielonka's algorithm in quasi-polynomial time. In: MFCS, pp. 10:1–10:13 (2019). https://doi.org/10.4230/LIPIcs.MFCS.2019.10

14. Parys, P.: Parity games: another view on Lehtinen's algorithm. In: CSL, pp. 32:1–32:15 (2020). https://doi.org/10.4230/LIPIcs.CSL.2020.32

15. Zielonka, W.: Infinite games on finitely coloured graphs with applications to automata on infinite trees. Theor. Comput. Sci. **200**(1–2), 135–183 (1998). https://doi.org/10.1016/S0304-3975(98)00009-7

Approximate Coalgebra Homomorphisms and Approximate Solutions

Jiří Adámek[1,2]([✉])

[1] Faculty of Electrical Engineering, Technical University Prague,
Prague, Czech Republic
[2] Institute of Theoretical Computer Science, Technical University Braunschweig,
Braunschweig, Germany
j.adamek@tu-bs.de

Abstract. Terminal coalgebras νF of finitary endofunctors F on categories called strongly lfp are proved to carry a canonical ultrametric on their underlying sets. The subspace formed by the initial algebra μF has the property that for every coalgebra A we obtain its unique homomorphism into νF as a limit of a Cauchy sequence of morphisms into μF called approximate homomorphisms. The concept of a strongly lfp category includes categories of sets, posets, vector spaces, boolean algebras, and many others.

For the free completely iterative algebra ΨB on a pointed object B we analogously present a canonical ultrametric on its underlying set. The subspace formed by the free algebra ΦB on B has the property that for every recursive equation in ΨB we obtain the unique solution as a limit of a Cauchy sequence of morphisms into ΦB called approximate solutions. A completely analogous result holds for the free iterative algebra RB on B.

1 Introduction

A number of important types of state-based systems can be expressed as coalgebras for finitary endofunctors F of **Set**. That is, endofunctors such that every element of FX lies in the image of Fm for some finite subset $m\colon M \hookrightarrow X$. Examples include deterministic automata, dynamic systems and finitely branching LTS. The terminal coalgebra νF carries a canonical ultrametric, see [2] or [4]. Recall from [12] that νF represents the set of all possible state behaviors, and given a system as a coalgebra A for F, the unique homomorphism $h\colon A \to \nu F$ assigns to every state its behavior. We are going to prove that whenever F is grounded, i.e., $F\emptyset \neq \emptyset$, then the initial algebra μF forms a dense subspace of the ultrametric space νF. And the coalgebra structure of νF is determined as the unique continuous extension of the inverted algebra structure of μF. Moreover, for every coalgebra A we present a canonical Cauchy sequence of morphisms

J. Adámek—Supported by the Grant Agency of the Czech Republic under the grant 19-00902S.

$h_k \colon A \to \mu F$ $(k < \omega)$ converging to h in the power space $(\nu F)^A$. We thus see that given state-based systems represented by a finitary set functor, (infinite) behaviour of states has canonical finite approximations, see Proposition 18.

All this generalizes from **Set** to every strongly locally finitely presentable category. These categories, introduced in [3] and recalled in Sect. 2 below, include vector spaces, posets and all locally finite varieties (e.g., boolean algebras or $\mathbb{S}$-**Mod**, semimodules over finite semirings $\mathbb{S}$). Every such category has a natural forgetful functor U into **Set**. And the underlying set $U(\nu F)$ of the terminal coalgebra νF carries a canonical ultrametric for every finitary endofunctor preserving monomorphisms. For example, consider the endofunctor of $\mathbb{S}$-**Mod** representing weighted automata, $FX = \mathbb{S} \times X^\Sigma$ (where Σ is a finite set of inputs). It has the terminal coalgebra $\nu F = \mathbb{S}^{\Sigma^*}$, the semimodule of weighted languages. The ultrametric assigns to distinct languages L_1, L_2 the distance 2^{-k} for the least k such that for some word $w \in \Sigma^*$ of length k the two weights are different, see Example 13(4).

If F is grounded, we again define, for every coalgebra A, a canonical sequence of morphisms $h_k \colon A \to \mu F$ converging to the unique coalgebra homomorphism $h \colon A \to \nu F$ in the power of the above ultrametric space. A concrete example: for every object Σ consider the endofunctor $FX = X \times \Sigma + 1$ where 1 is the terminal object. The underlying set of νF is the set of all finite and infinite words over $U\Sigma$. And in case U preserves finite coproducts, the underlying set of μF is the set of all finite words. The distance of two distinct words is 2^{-k} for the least k such that their prefixes of lengths at most k are distinct.

Recursive equations in the terminal coalgebra (considered as an algebra) have a unique solution. That is, the algebra νF is completely iterative, see [11]. We define, analogously, a canonical Cauchy sequence of approximate solutions into the initial algebra. And we prove that this sequence converges to the unique solution in νF. We recall from Milius [11] the concept of an algebra A for F being completely iterative in Sect. 4. Shortly, this means that for every object X (of recursion variables) and every equation morphism $e \colon X \to FX + A$ there exists a unique solution, which is a morphism $e^\dagger \colon X \to A$ satisfying a simple recursive condition. Milius proved that the free completely iterative algebra ΨB on an object B is precisely the terminal coalgebra for the endofunctor $F(-) + B$:

$$\Psi B = \nu X. FX + B.$$

The functor $F(-) + B$ is grounded whenever B has a global element. Thus we have the canonical ultrametric on the underlying set of ΨB. Moreover, the free algebra ΦB for F on B is well-known to be the initial algebra for the functor $F(-) + B$:

$$\Phi B = \mu X. FX + B.$$

We are going to present, for every recursive equation e in ΨB, a canonical Cauchy sequence of 'approximate solutions' $e_k^\dagger \colon X \to \Phi B$ in the free algebra which converges to the solution $e^\dagger$ in the power space of νF, see Theorem 31.

Finally, the same result is proved for the free iterative algebra RB: solutions are limits of sequences of approximate solutions. Recall that 'iterative' is the weakening of 'completely iterative' by restricting the objects X (of recursion variables) to finitely presentable ones.

Related Work. For finitary set functors preserving ω^{op}-limits Barr proved that the canonical ultrametric of νF is complete, and in case F is grounded this is the Cauchy completion of the subspace μF, see [10]. This was generalized to strongly lfp categories in [3]. The fact that preservation of ω^{op}-limits is not needed is new, and so is the fact that the algebra structure of μF determines the coalgebra structure of νF.

The idea of approximate homomorphisms and approximate solutions has been used for set functors in the recent paper [5] where instead of ultrametrics complete partial orders were applied. It turns out that the present approach is much more natural and needs less technical preparations.

2   Strongly lfp Categories

Throughout the paper we work with a finitary endofunctor F of a locally finitely presentable (shortly lfp) category. In this preliminary section we recall basic properties of lfp categories and introduce strongly lfp categories.

Recall that an object A of a category $\mathcal{K}$ is called *finitely presentable* if its hom-functor $\mathcal{K}(A, -)$ preserves filtered colimits. And a category $\mathcal{K}$ is called lfp, or *locally finitely presentable*, if it has colimits and a set of finitely presentable objects whose closure under filtered colimits is all of $\mathcal{K}$. Every lfp category is complete ([8], Remark 1.56).

Recall further that an endofunctor is called *finitary* if it preserves filtered colimits.

Example 1 (See [6]). *(1)* **Set** *is lfp. An endofunctor F is finitary iff every element of FX lies in the image of Fm for some finite subset $m\colon M \hookrightarrow X$.*

(2) K-**Vec**, *the category of vector space over a field K, is lfp. An endofunctor F is finitary iff every element of FX lies in the image of Fm for some finite-dimensional subspace $m\colon M \hookrightarrow X$.*

Remark 1. (1) In an lfp category, given a filtered colimit $a_t\colon A_t \to A$ ($t \in T$) of a diagram D with all connecting morphisms monic, it follows that

(a) each a_t is monic, and

(b) for every cocone $f_t\colon A_t \to X$ of monomorphisms for D the unique factorization morphism $f\colon A \to X$ is monic.

See [8], Proposition 1.62.

(2) In *any* category the corresponding statement is true for limits $a_t\colon A \to A_t, t < \lambda$, of λ-chains of monomorphisms for all limit ordinals λ:

(a) each a_t is monic, and

(b) for every cone $f_t\colon X \to A_t$ of monomorphisms the factorization morphism $f\colon X \to A$ is monic.

Indeed, (a) follows from the fact that the cone of all a_s with $t \leq s$ is collectively monic. And (b) is trivial: since $a_0 \cdot f$ is monic, so is f.

(3) Every lfp category has (strong epi, mono)-factorizations of morphisms, see [8], Proposition 1.61. An epimorphism $e\colon A \to B$ is called *strong* if for every monomorphism $m\colon U \to V$ the following diagonal fill-in property holds: given a commutative square $v \cdot e = m \cdot u$ (for some $u\colon A \to U$ and $v\colon B \to V$) there exists a unique 'diagonal' morphism $d\colon B \to U$ with $u = d \cdot e$ and $v = m \cdot d$.

(4) An object A of an lfp category $\mathcal{K}$ is called *finitely generated* if it is a strong quotient of a finitely presentable object. Equivalently: $\mathcal{K}(A, -)$ preserves filtered colimits of monomorphisms, see [8], Proposition 1.69.

(5) The initial object of $\mathcal{K}$ is denoted by 0. We call it *simple* if it has no proper strong quotients or, equivalently, if the unique morphism from it to any given object is monic. This holds e.g. in **Set** and K-**Vec**. In the lfp category $\mathbf{Alg}_{1,0}$ of unary algebras with a constant the intial object is the algebra of natural numbers with the successor operation, and it is not simple.

Notation 2. *(1) We denote by $F^n 0$ ($n < \omega$) the initial-algebra ω-chain*

$$0 \xrightarrow{\;!\;} F0 \xrightarrow{\;F!\;} F^2 0 \xrightarrow{\;F^2!\;} \cdots$$

with connecting maps $w_{n,k}\colon F^n 0 \to F^k 0$, where $w_{0,1} =\,!$ is unique and $w_{n+1,k+1} = F w_{n,k}$. The colimit cocone $w_{n,\omega}\colon F^n 0 \to F^\omega 0$ ($n < \omega$) of this chain yields, since F is finitary, a colimit cocone $F w_{n,\omega}\colon F^{n+1} 0 \to F(F^\omega 0)$. Consequently, we obtain a unique morphism

$$\varphi\colon F(F^\omega 0) \to F^\omega 0 \quad \text{with} \quad \varphi \cdot F w_{n,\omega} = w_{n+1,\omega} \quad (n < \omega).$$

Then $F^\omega 0$ is the initial algebra with the algebra structure φ, see the second proposition of [1].

(2) We denote by $F^i 1$ ($i \in \mathrm{Ord}$) the terminal-coalgebra chain indexed by all ordinals, dually ordered:

$$1 \xleftarrow{\;!\;} F1 \xleftarrow{\;F!\;} FF1 \xleftarrow{\;F^2!\;} \cdots \quad F^i 1 \leftarrow \quad \cdots$$

It is given by transfinite recursion: $F^0 1 = 1$, $F^{i+1} 1 = F(F^i 1)$ and, for limit ordinals i, $F^i 1 = \lim_{j<i} F^j 1$. The connecting maps $v_{ij}\colon F^i 1 \to F^j 1$ (for $i \geq j$ in Ord) are determined by transfinite recursion as follows:

$$v_{1,0} =\,!\colon F1 \to 1 \quad (unique), \qquad v_{i+1,j+1} = F v_{i,j}$$

and for limit ordinals i the above limit $F^i 1$ has limit projections v_{ij} ($j < i$).

(3) In case F preserves limits of ω^{op}-sequences the morphism $v_{\omega+1,\omega}$ is invertible, thus

$$\nu F = F^\omega 1,$$

with the coalgebra structure given by $\tau = v_{\omega+1,\omega}^{-1}$.

Theorem 3 ([9], **Thm. 2.6**). *If a finitary endofunctor F of an lfp category preserves monomorphisms, then it has a terminal coalgebra $\nu F = F^i 1$ for some ordinal $i \geq \omega$ with $v_{i+1,i}$ invertible. The coalgebra structure is $\tau = v_{i+1,i}^{-1} \colon F^i 1 \to F(F^i 1)$.*

Proposition 4. *Let $\mathcal{K}$ be an lfp category with a simple initial object. For every finitary endofunctor F preserving monomorphisms the initial algebra is a canonical subobject of the terminal coalgebra: the unique algebra homomorphism*

$$m \colon (\mu F, \varphi) \to (\nu F, \tau^{-1})$$

is monic.

Proof. The initial-algebra ω-chain has a unique cocone $m_k \colon F^k 0 \to \nu F$ ($k < \omega$) satisfying the following recursion

$$m_{k+1} \equiv F(F^k 0) \xrightarrow{Fm_k} F(\nu F) \xrightarrow{\tau^{-1}} \nu F \quad (k < \omega).$$

Since $m_0 \colon 0 \to \nu F$ is monic and F preserves monomophisms, each m_k is monic. Therefore, the unique morphism

$$m \colon F^\omega 0 \to \nu F \quad \text{with} \quad m \cdot w_{k,\omega} = m_k \quad (k < \omega)$$

is monic, see Remark 1(1). And this is an algebra homomorphism, i.e., $m \cdot \varphi = \tau^{-1} \cdot Fm \colon F(\mu F) \to \nu F$. To verify this, recall that F preserves colimits of chains, hence $(Fw_{k,\omega})_{k<\omega}$ is a collectively epic cocone. Thus, we only need to verify

$$m \cdot \varphi \cdot Fw_{k,\omega} = \tau^{-1} \cdot Fm \cdot Fw_{k,\omega} \quad (k < \omega).$$

By definition of φ, the left-hand side is $m \cdot w_{k+1,\omega} = m_{k+1}$. The right-hand one is the same: $\tau^{-1} \cdot Fm_k = m_{k+1}$.

Remark 2. An object G of $\mathcal{K}$ is called a *generator* if its hom-functor $U = \mathcal{K}(G, -)$ is faithful. Thus $\mathcal{K}$ becomes a concrete category w.r.t. that hom-functor. For example K is a generator of K-**Vec**, and the terminal object 1 is a generator of **Set** or **Pos** (the category of posets). In all these examples the above functor U is naturally isomorphic to the usual forgetful functor.

Definition 5 *(See [4]).* An lfp category is called strongly lfp provided that
(a) for every limit cone $a_k \colon A \to A_k$ ($k < \omega$) of an ω^{op}-chain and every monomorphism $m \colon M \to A$ with M finitely generated there exists k with $a_k \cdot m \colon M \to A_k$ monic, and
(b) a generator G is given.

In [4] the generator was assumed to be projective, but we do not need this assumption. Moreover, in that paper instead of our present factorization system (strong epi, mono) the system (epi, strong mono) was used.

Example 6. *(1)* **Set** *is strongly lfp. Indeed, given an ω^{op}-limit $a_k\colon A \to A_k$ ($k < \omega$) and a finite subset $m\colon M \to A$, for every pair $x \neq y$ in M there exists k with $a_k(x) \neq a_k(y)$. Since $M \times M$ is finite, k can be chosen independent of x, y. Thus, $a_k \cdot m$ is monic.*

(2) Every variety of algebras which is locally finite (i.e., finitely generated algebras are finite) is strongly lfp. The free algebra G on one generator is a generator. And since limits are formed on the level of sets, the ω^{op}-limit condition is verified precisely as in (1).

*Examples include semilattices, boolean algebras, categories $\mathbb{S}$-**Mod** of left semimodules over a finite semiring $\mathbb{S}$, and categories M-**Set** of sets with an action of a finite monoid M.*

*(3) The category K-**Vec** of vector spaces is strongly lfp for every (possibly infinite) field K, see [3], Example 4.5(5).*

*(4) The category **Pos** of posets is strongly lfp whose terminal object is a generator. The ω^{op}-limit condition is verified as in (1) above.*

*(5) The category **Gra** of graphs (sets with a binary relation) and graph homomorphisms is strongly lfp. The single-vertex graph G without edges is a generator, and the ω^{op}-limit condition is verified as in (1).*

Notation 7. *Given a strongly lfp category $\mathcal{K}$ with a generator G, we denote by*

$$U\colon \mathcal{K} \to \mathbf{Set}$$

the hom-functor of G. This is a faithful functor preserving limits (in particular, it preserves monomorphisms).

Remark 3. (1) In contrast to examples (2) and (3) above, the categories $\mathbb{Z}$-**Mod** of abelian groups and $\mathbb{Z}$-**Set** of unary algebras on one operation are not strongly lfp, see [3], Example 4.5.

(2) Worrell [13] proved that for every finitary set functor F all the connecting maps $v_{i,\omega}$, $i > \omega$, are monic. This can be generalized as follows.

Theorem 8. *Let $\mathcal{K}$ be a strongly lfp category. For every finitary endofunctor F preserving monomorphisms the terminal coalgebra $\nu F = F^i 1$ (see Theorem 3) is a canonical subobject of $F^\omega 1$: the morphism $v_{i\omega}$ is monic.*

Proof. We prove that for all ordinals $j > \omega$ the connecting maps $v_{j\omega}$ are monic by transfinite induction.

(1) $v_{\omega+1,\omega}\colon F(F^\omega 1) \to F^\omega 1$ is monic. To prove this, use the fact that $\mathcal{K}$ is lfp and choose a filtered diagram D of finitely presentable objects A_t ($t \in T$) with a colimit cocone $a_t\colon A_t \to F^\omega 1$. Due to Remark 1(3) we have a factorization $a_t = b_t \cdot e_t$ for a strong epimorphism $e_t\colon A_t \to B_t$ and a monomorphism $b_t\colon B_t \to F^\omega 1$. The objects B_t ($t \in T$) form a filtered diagram $\bar{D}$ whose connecting morphisms are derived via diagonal fill-in: given a connecting morphism $a_{t,t'}\colon A_t \to A_{t'}$

in D, we obtain a connecting morphism $b_{t,t'} : B_t \to B_{t'}$ of $\bar{D}$ by means of the following diagonal:

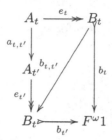

It is easy to verify that $\bar{D}$ a filtered diagram (since D is) and the morphisms $b_t : B_t \to F^\omega 1$ $(t \in T)$ form the colimit cocone of $\bar{D}$. Moreover, each connecting morphism $b_{t,t'}$ is monic (since its composite with $b_{t'}$ is). Hence, our assumption that F is finitary and preserves monomorphisms implies that we have a filtered diagram $F\bar{D}$ of all FB_t $(t \in T)$ whose connecting morphisms are monic and whose colimit is $(Fb_t)_{t \in T}$. Using Remark 1(1) it is sufficient, for proving that $v_{\omega+1,\omega}$ is monic, to verify that (for every $t \in T$)

$$v_{\omega+1,\omega} \cdot Fb_t : FB_t \rightarrowtail F^\omega 1 \quad \text{is monic.}$$

Since our category is strongly lfp and B_t is finitely generated (being a strong quotient of the finitely presentable object A_t), it follows that for the monomorphism b_t there exists $k < \omega$ with $v_{\omega,k} \cdot b_t$ monic. Thus, the morphism $Fv_{\omega,k} \cdot Fb_t = v_{\omega,k+1} \cdot v_{\omega+1,\omega} \cdot Fb_t = v_{\omega+1,\omega} \cdot Fb_t$ is monic, as desired.

(2) Each $v_{i,\omega}$ for $i > \omega$ is monic. This is easily seen by transfinite induction on i. The case $i = \omega+1$ has just been proved. If $v_{i,\omega}$ is monic, so is $Fv_{i,\omega} = v_{i+1,\omega+1}$, therefore, the composite $v_{i+1,\omega} = v_{\omega+1,\omega} \cdot v_{i+1,\omega+1}$ is also monic. And given a limit ordinal $i > \omega$, then each connecting map $v_{k,j}$, where $\omega \leq j \leq k < i$ is monic, since $v_{j,\omega} \cdot v_{k,j} = v_{k,\omega}$ is monic, see Remark 1(2). Now $F^i 1$ is a limit of the chain $F^j 1$ for all $\omega \leq j < i$ whose connecting morphisms are monic. This implies that the limit cone consists of monomorphisms. In particular, $v_{i,\omega}$ is monic, as required.

Lemma 9. *For i in Theorem 8 the following squares commute:*

$$
\begin{array}{ccc}
\mu F & \xrightarrow{\;\;m\;\;} & \nu F = F^i 1 \qquad (k < \omega) \\
{\scriptstyle w_{k,\omega}} \uparrow & & \downarrow {\scriptstyle v_{i,k}} \\
F^k 0 & \xrightarrow[F^k !]{} & F^k 1
\end{array}
$$

Proof. We proceed by induction, the case $k = 0$ is clear. Recall that m is an algebra homomorphism:

$$m \cdot \varphi = \tau^{-1} \cdot Fm = v_{i+1,i} \cdot Fm. \tag{2.1}$$

By definition of φ we have $\varphi \cdot Fw_{k,\omega} = w_{k+1,\omega}$ (Notation 2(1)), thus,

$$m \cdot w_{k+1,\omega} = v_{i+1,i} \cdot Fm \cdot Fw_{k,\omega}. \tag{2.2}$$

Assuming the above square commutes for k, we prove it for $k + 1$:

$$
\begin{aligned}
F^{k+1}! &= F(v_{i,k} \cdot m \cdot w_{k,\omega}) && \text{induction hypothesis} \\
&= v_{i+1,k+1} \cdot Fm \cdot Fw_{k,\omega} && v_{i+1,k+1} = Fv_{i,k} \\
&= v_{i,k+1} \cdot v_{i+1,i} \cdot Fm \cdot Fw_{k,\omega} && \\
&= v_{i,k+1} \cdot m \cdot w_{k+1,\omega} && \text{by (2.2)}
\end{aligned}
$$

Since $v_{i,k}$ is the composite $v_{\omega,k} \cdot v_{i,\omega}$, we obtain the following corollary of Theorem 8:

Corollary 10. *If the initial object is simple, then the following monomorphism*

$$\bar{u} \equiv \mu F \xrightarrow{\ m\ } F^i 1 \xrightarrow{\ v_{i,\omega}\ } F^\omega 1$$

makes the squares below commutative

$$
\begin{array}{ccc}
\mu F & \xrightarrow{\ \bar{u}\ } & F^k 1 \qquad (k < \omega) \\
{\scriptstyle w_{k,\omega}}\big\uparrow & & \big\downarrow{\scriptstyle v_{i,k}} \\
F^k 0 & \xrightarrow[\ F^k!\ ]{} & F^k 1
\end{array}
$$

3   Approximate Homomorphisms

Assumption 11. *Throughout the rest of the paper $\mathcal{K}$ denotes a strongly lfp category with a simple initial object, and F a finitary endofunctor preserving monomorphisms.*

Recall that the initial algebra μF is a canonical subobject of the terminal coalgebra νF (Proposition 4). For the forgetful functor of Notation 7 we present a canonical ultrametric on the underlying set $U(\nu F)$ of the terminal coalgebra. Then $U(\mu F)$ is a subspace of that ultrametric space, since U preserves monomorphisms. We prove that those two ultrametric spaces have the same Cauchy completion. Moreover, given an arbitrary coalgebra $\alpha\colon A \to FA$, the unique homomorphism into νF is determined by a Cauchy sequence of (approximate) morphisms from A to μF.

Remark 4. (1) Recall that a metric d is called an *ultrametric* if the triangle inequality can be strengthened to $d(x,z) \leq \max\big(d(x,y), d(y,z)\big)$. We denote by **UMet** the category of ultrametric space with distances at most 1 and non-expanding functions. That is, functions $f\colon (X,d) \to (X',d')$ with $d(x,y) \geq d'\big(f(x), f(y)\big)$ for all $x, y \in X$.

(2) The category **UMet** has products w.r.t. the supremum metric. In particular, for every set M the power $(X, d)^M$ is the ultrametric space of all functions $f \colon M \to X$ with distances given by

$$d(f, f') = \sup_{m \in M} \big(d(f(m), f'(m)) \big) \quad \text{for} \quad f, f' \in X^M.$$

(3) Completeness of an ultrametric space means that every Cauchy sequence has a limit. Every ω^{op}-sequence of morphisms $a_{k+1,k} \colon A_{k+1} \to A_k$ $(k < \omega)$ in **Set** carries a canonical complete ultrametric on its limit $A = \lim_{k < \omega} A_k$ in **Set** (with limit projections $a_k \colon A \to A_k$)): Given elements $x \neq y$ of A, their distance is defined by

$$d(x, y) = 2^{-k}$$

for the least k with $a_k(x) \neq a_k(y)$. It is easy to verify that (A, d) is a complete ultrametric space, see also [4], Lemma 32.

(4) We apply this to the limit $F^\omega 1 = \lim_{n < \omega} F^n 1$ of Notation 2(2). Since the forgetful functor $U = \mathcal{K}(G, -)$ preserves limits, $U(F^\omega 1)$ is a limit with projections $U v_{\omega, k}$ $(k < \omega)$.

Corollary 12. *The underlying set of $F^\omega 1$ carries the following complete ultrametric: given distinct elements x, $y \colon G \to F^\omega 1$ of $U(F^\omega 1)$, their distance is 2^{-k} for the least k with $v_{\omega, k} \cdot x \neq v_{\omega, k} \cdot y$.*

The monomorphisms

$$m \colon \mu F \to \nu F \quad \text{and} \quad v_{i,\omega} \colon \nu F \to F^\omega 1$$

of Proposition 4 and Theorem 3 are preserved by U, thus we have ultrametric subspaces $U(\mu F) \subseteq U(\nu F) \subseteq U(F^\omega 1)$.

Example 13. *(1) For the set functor $FX = \Sigma \times X + 1$ (whose coalgebras are dynamic systems with inputs from Σ and terminating states) the terminal-coalgebra chain*

$$1 \leftarrow \Sigma + 1 \leftarrow \Sigma \times \Sigma + \Sigma + 1 \leftarrow \cdots$$

is given by $F^k 1 = $ words of length $\leq k$. And $v_{k+1, k}$ deletes the first letter in every word of length $k + 1$. The limit of this ω^{op}-sequence is

$$\nu F = F^\omega 1 = \Sigma^\omega + \Sigma^*.$$

The projection $v_{\omega, k}$ leaves words of length $\leq k$ unchanged and otherwise forms prefixes of length k. Thus the distance of distinct words is 2^{-k} for the least k for which their prefixes of length at most k are distinct.

The initial-algebra chain

$$0 \to 1 \to \Sigma + 1 \to (\Sigma \times \Sigma) + \Sigma + 1 \to \cdots$$

is given by $F^k0 =$ words of length less than k. And $w_{k,k+1}$ is the inclusion. This yields the colimit

$$\mu F = \Sigma^*$$

with the above ultrametric.

(2) More generally, given a set B, then $FX = \Sigma \times X + B$ has the terminal coalgebra $\nu F = \Sigma^\omega + \Sigma^* \times B$. Words in $\Sigma^* \times B$ with distinct B-components have distance 1, otherwise the ultrametric is analogous to (1). And $\mu F = \Sigma^* \times B$.

(3) Deterministic automata with the input alphabet Σ are coalgebras of the set functor $FX = \{0,1\} \times X^\Sigma$. The terminal-coalgebra chain

$$1 \leftarrow \{0,1\} \leftarrow \{0,1\}^{1+\Sigma} \leftarrow \{0,1\}^{1+\Sigma+\Sigma^2} \leftarrow \cdots$$

is given by powers of $\{0,1\}$ to $\coprod_{n=0}^{k} \Sigma^n$. Thus, F^k1 are all formal languages using words of length at most k. And $v_{k+1,k}$ deletes the first letter in every word of length $k+1$. Since F preserves limits of ω^{op}-sequences, we have

$$\nu F = F^\omega 1 = \text{all formal languages.}$$

The distance of distinct languages is 2^{-k} for the least k such that there is a word of length k lying in precisely one of those languages. Here $\mu F = \emptyset$.

*(4) Weighted automata with weights in a finite semiring $\mathbb{S}$ and the input alphabet Σ are coalgebras for the endofunctor of $\mathbb{S}$-**Mod** defined by $FX = \mathbb{S} \times X^\Sigma$. The terminal-coalgebra chain is given by left-hand projections π_1:*

$$1 \xleftarrow{\ \pi_1=0\ } \mathbb{S} \xleftarrow{\ \pi_1\ } \mathbb{S} \times \mathbb{S}^\Sigma \xleftarrow{\ \pi_1\ } (\mathbb{S} \times \mathbb{S}^\Sigma) \times \mathbb{S}^{\Sigma \times \Sigma} \simeq \mathbb{S}^{1+\Sigma+\Sigma^2} \xleftarrow{\ \pi_1\ } \cdots$$

This yields as F^k1 the set of all weighted languages with weight 0 for all words longer than k. And $v_{k+1,k}$ changes weights of words of length $k+1$ to 0. Here, again, F preserves limits of ω^{op}-sequences, thus

$$\nu F = F^\omega 1 = \mathbb{S}^{\Sigma^*}$$

the set of all weighted languages. The distance of distinct languages is 2^{-k} for the least k with their k-components distinct.

*The initial-algebra chain has the same objects, and $w_{k,k+1}$ are the coproduct injections (using that in $\mathbb{S}$-**Mod** we have $A + B = A \times B$). Thus the initial algebra is*

$$\mu F = \text{all bounded weighted languages,}$$

where bounded means that for some number k all words of nonzero weight have length at most k.

(5) Let Γ be a signature, $\Gamma = (\Gamma_n)_{n \in \mathbb{N}}$. The polynomial set functor $H_\Gamma X = \coprod_{n \in \mathbb{N}} \Gamma_n \times X^n$ has the terminal coalgebra νH_Γ consisting of all Γ-trees, i.e., trees

labelled by Γ so that a node with a label in Γ_n has n successors. (We always consider trees up to isomorphism.) We can represent $H_\Gamma^k 0$ by the set of all Γ-trees of height less than k, thus the colimit

$$\mu H_\Sigma = \bigcup_{k \in \mathbb{N}} H_\Gamma^k 0$$

is given by all finite Γ-trees.

The distance of trees $t \neq s$ in νH_Γ is 2^{-k} for the least k such that by cutting t and s at level k one obtains distinct trees.

Remark 5. Example (1) above generalizes to all strongly lfp categories whose forgetful functor $U = \mathcal{K}(G, -)$ preserves finite coproducts, e.g. **Pos** and **Gra**. Given an object Σ, the functor $FX = X \times \Sigma + 1$ is finitary and the forgetful functor U (preserving all limits) takes the terminal-coalgebra chain

$$1 \leftarrow \Sigma + 1 \leftarrow \Sigma \times \Sigma + \Sigma + 1 \leftarrow \cdots$$

in $\mathcal{K}$ to the terminal-coalgebra chain of the set-functor $(-) \times U\Sigma + 1$. The terminal coalgebra νF has the underlying set $U(\nu F) = (U\Sigma)^\omega + (U\Sigma)^*$, and the distance of words is as described above. If, moreover, G is a finitely presentable generator (which is true in all of our examples), then U also takes the initial-algebra chain of F to that of the set functor $(-) \times U\Sigma + 1$. Thus, $U(\mu F) = (U\Sigma)^*$.

Definition 14. *An endofunctor F is grounded if $F0$ has a global element, i.e., a morphism $p \colon 1 \to F0$ is given.*

Proposition 15. *The ultrametric subspace $U(\mu F)$ is dense in $U(\nu F)$ whenever F is grounded. These two spaces have the same Cauchy completion, viz, $U(F^\omega 1)$.*

Proof. (1) We prove that the embedding $U\bar{u} \colon U(\mu F) \to U(F^\omega 1)$, see Corollary 10, is dense. Since $U(F^\omega 1)$ is complete, both statements above follow from this fact. Thus given an element $x \colon G \to F^\omega 1$ of $U(F^\omega 1)$, we are to present a sequence $x_k \colon G \to F^\omega 1$ $(k < \omega)$ of morphisms factorizing through $\bar{u}$ with $x = \lim_{k \to \infty} x_k$ in the supremum metric of Remark 4(2).

(2) For every $k < \omega$ consider the following endomorphism of $F^\omega 1$ using the global element $p \colon 1 \to F0$:

$$r_k \equiv F^\omega 1 \xrightarrow{v_{\omega,k}} F^k 1 \xrightarrow{F^k p} F^{k+1} 0 \xrightarrow{w_{k+1,\omega}} F^\omega 0 \xrightarrow{\bar{u}} F^\omega 1. \tag{3.1}$$

It fulfils, due to Corollary 10

$$v_{\omega,k+1} \cdot r_k = F^{k+1}! \cdot F^k p \cdot v_{\omega,k}. \tag{3.2}$$

In Notation 2 we see that $v_{k+1,k} = F^k v_{1,0}$, thus the (obvious) equality $v_{1,0} \cdot F! \cdot p = \mathrm{id}_1$ yields

$$v_{k+1,k} \cdot F^{k+1}! \cdot F^k p = \mathrm{id}_{F^k 1}. \tag{3.3}$$

Consequently

$$v_{\omega,k} \cdot r_k = v_{\omega,k}. \tag{3.4}$$

Indeed, by (3.2) and $v_{\omega,k} = v_{k+1,k} \cdot v_{\omega,k+1}$ we see that the left-hand side is $v_{k+1,k} \cdot F^{k+1}! \cdot F^k p \cdot v_{\omega,k}$ which is $v_{\omega,k}$ by (3.3).

(3) The desired sequence is

$$x_k = r_k \cdot x.$$

It converges to x because (3.4) yields $v_{\omega,k} \cdot (r_k \cdot x) = v_{\omega,k} \cdot x$, thus $d(r_k \cdot x, x) < 2^{-k}$. And since r_k factorizes through $\bar{u}$, this concludes the proof.

The following corollary was already proved by Barr [10]:

Corollary 16. *If, moreover, F preserves limits of ω^{op} sequences, then the ultra-metric space corresponding to νF is the Cauchy completion of the subspace μF.*

For every coalgebra $\alpha \colon A \to FA$ carried by a set $M = UA$ the unique coalgebra homomorphism $h \colon A \to \nu F$ lies in the metric space $U(\nu F)^M$, see Remark 4(2). And $U(\mu F)^M$ is a subspace via the post-composition with m of Proposition 4. We are going to prove that h is the limit of a canonical Cauchy sequence in the subspace $U(\mu F)^M$.

Definition 17. *For every coalgebra (A, α) we define the sequence $h_k \colon A \to \mu F$ $(k < \omega)$ of approximate homomorphisms by induction as follows*

$$h_0 \equiv A \xrightarrow{\ !\ } 1 \xrightarrow{\ p\ } F0 \xrightarrow{\ w_{1,\omega}\ } \mu F$$

and

$$h_{k+1} \equiv A \xrightarrow{\ \alpha\ } FA \xrightarrow{\ Fh_k\ } F(\mu F) \xrightarrow{\ \varphi\ } \mu F. \tag{3.5}$$

Thus h_0 is a trivial approximation, independent of the given coalgebra structure: it just uses the global element p. Given an approximation h_k, the next one is given by a square similar to the square defining coalagebra homomorphisms:

$$
\begin{array}{ccc}
A & \xrightarrow{\ h_{k+1}\ } & \mu F \\
{\scriptstyle \alpha}\downarrow & & \downarrow{\scriptstyle \varphi^{-1}} \\
FA & \xrightarrow[\ Fh_k\ ]{} & F(\mu F)
\end{array}
$$

Let us remark that we do not claim that h_k are coalgebra homomorphisms.

Proposition 18. *For every coalgebra (A, α) on the set $M = UA$ the unique coalgebra homomorphism $h \colon A \to \nu F$ is a limit of the approximate homomor-phisms. More precisely: in the ultrametric space $U(\nu F)^M$ we have*

$$Uh = \lim_{k \to \infty} U(m \cdot h_k).$$

Proof. We prove by induction on k that $d\big(Uh, U(m \cdot h_k)\big) \leq 2^{-k}$. This is trivial for $k = 0$ (recall that we assume that all distances are at most 1). If this holds for k, we prove it for $k + 1$. Since h is a coalgebra homomorphism and we have an ordinal i with $\tau^{-1} = v_{i+1,i}$ (see Theorem 3), we get

$$h = v_{i+1,i} \cdot Fh \cdot \alpha. \tag{3.6}$$

As m is an algebra homomorphism (see Proposition 4), we have

$$m \cdot \varphi = \tau^{-1} \cdot Fm = v_{i+1,i} \cdot Fm \tag{3.7}$$

The induction hypothesis is that $v_{i,k-1}$ merges h and $m \cdot h_k$:

$$v_{i,k-1} \cdot h = v_{i,k-1} \cdot m \cdot h_k \tag{3.8}$$

and our task is to prove that $v_{i,k}$ merges h and $m \cdot h_{k+1}$:

$$
\begin{aligned}
v_{i,k} \cdot h &= v_{i,k} \cdot v_{i+1,i} \cdot Fh \cdot \alpha && \text{by (3.6)} \\
&= Fv_{i,k-1} \cdot Fh \cdot \alpha && v_{i,k} \cdot v_{i+1,i} = v_{i+1,k} = Fv_{i,k-1} \\
&= F\big(v_{i,k-1} \cdot m \cdot h_k\big) \cdot \alpha && \text{by (3.8)} \\
&= v_{i,k} \cdot v_{i+1,i} \cdot Fm \cdot Fh_k \cdot \alpha && \text{as above} \\
&= v_{i,k} \cdot m \cdot \varphi \cdot Fh_k \cdot \alpha && \text{by (3.7)} \\
&= v_{i,k} \cdot m \cdot h_{k+1} && \text{by (3.5)}
\end{aligned}
$$

Notation 19. *Define morphisms $\partial_k \colon \nu F \to \mu F$ for $k < \omega$ as the following composites*

$$\partial_k \equiv F^i 1 \xrightarrow{v_{i,k}} F^k 1 \xrightarrow{F^k p} F^{k+1} 0 \xrightarrow{w_{k+1,\omega}} F^\omega 0.$$

We prove below that this sequence (prolonged by the embedding $m \colon \mu F \to \nu F$) converges to $\mathrm{id}_{\nu F}$. The following examples give an intuition how these approximations of the identity morphism work.

Example 20. *(1) For $FX = X \times \Sigma + B$ of Example 13(2) with a chosen element $p \in B$ the map $\partial_k \colon \Sigma^\omega + \Sigma^* \times B \to \Sigma^* \times B$ assigns to an infinite word w the pair (w', p) where w' is the prefix of length k. And to pairs $(w, b) \in \Sigma^* \times B$ it assigns analogously (w', p) if $|w| \geq k$, whereas $\partial_k(w, b) = (w, b)$ if $|w| < k$.*

(2) For a polynomial functor H_Γ of Example 13(4) we have $\partial_k \colon \nu H_\Gamma \to \mu H_\Gamma$ which cuts every Γ-tree at height k and relabels all leaves of height k by p (the chosen nullary symbol).

Lemma 21. *For every $k < \omega$ we have $v_{i,k} = v_{i,k} \cdot m \cdot \partial_k$. Therefore, the sequence $m \cdot \partial_k$ converges to $\mathrm{id}_{\nu F}$ in the ultrametric space of endomorphisms of $U(\nu F)$:*

$$\lim_{k \to \infty} U(m \cdot \partial_k) = \mathrm{id}_{U(\nu F)} .$$

Proof. It is sufficient to prove that first identity, since this implies that $\mathrm{id}_{U(\nu F)}$ and $U(m \cdot \partial_k)$ have distance smaller than 2^{-k}.

From Notation 2 we know that $v_{i,k} = v_{k+1,k} \cdot v_{i,k+1}$. Thus

$$
\begin{aligned}
v_{i,k} \cdot m \cdot \partial_k &= v_{k+1,k} \cdot v_{i,k+1} \cdot m \cdot w_{k+1,\omega} \cdot F^k p \cdot v_{i,k} && \text{by def. of } \partial_k \\
&= v_{k+1,k} \cdot F^{k+1}! \cdot F^k p \cdot v_{i,k} && \text{Lemma 9} \\
&= F^k(v_{1,0} \cdot F! \cdot p) \cdot v_{i,k} && v_{k+1,k} = F v_{1,0}^k \\
&= v_{i,k} && v_{1,0} \cdot F! \cdot p = \mathrm{id}_1
\end{aligned}
$$

4 Approximate Solutions

We apply the above results to the finitary endofunctor

$$F_B = F(-) + B$$

for an arbitrary object B. We want F_B to preserve monomorphisms whenever F does. We therefore need the following (mild) additional assumption:

Assumption 22. *Additionally to the assumptions of Sect. 3 we assume that $\mathcal{K}$ has monic coproduct injections and that $m + B\colon X + B \to Y + B$ is monic for every monomorphism $m\colon X \to Y$. (This holds in all Examples of 6.)*

Notation 23. *Assume that B is a pointed object, i.e., a global element $p\colon 1 \to B$ is given. Then F_B is a grounded endofunctor w.r.t. the following global element*

$$p_B \equiv 1 \xrightarrow{\ p\ } B \xrightarrow{\ \mathrm{inr}\ } F0 + B = F_B 0.$$

Definition 24 (See [11]). *Let an algebra $\alpha\colon FA \to A$ be given. A (recursive) equation with an object X (of recursive variables) is a morphism $e\colon X \to FX + A$. A solution of e is a morphism $e^\dagger\colon X \to A$ making the following square*

$$
\begin{array}{ccc}
X & \xrightarrow{\ e^\dagger\ } & A \\
{\scriptstyle e}\big\downarrow & & \big\uparrow{\scriptstyle [\alpha, A]} \\
FX + A & \xrightarrow[\ Fe^\dagger + A\ ]{} & FA + A
\end{array}
$$

commutative.

An algebra in which every equation has a unique solution is said to be completely iterative.

Example 25 (see [11]). *(1) The terminal coalgebra νF considered as an algebra (via τ^{-1}) is completely iterative.*

(2) Denote by Ψ_B the terminal coalgebra of the endofunctor $F_B = F(-) + B$,

$$\Psi_B = \nu F_B.$$

Its coalgebra structure $\tau_B : \Psi B \to F(\Psi B) + B$ yields an inverse with components

$$\psi_B : F(\Psi B) \to \Psi B \quad and \quad \eta_B : B \to \Psi B.$$

The resulting algebra ΨB for F is completely iterative.

   Indeed, this is the free completely iterative algebra on B. That is, for every completely iterative algebra $\alpha : FA \to A$ and every morphism $f : B \to A$ there exists a unique algebra homomorphism $f^{\#} : \Psi B \to A$ with $f = f^{\#} \cdot \eta_B$.

   (3) Let us illustrate the iterativity of ΨB on the set functor $FX = \Sigma \times X$ where $\Sigma = \{\sigma, \tau\}$. From Example 13(2) we know that $\Psi B = \Sigma^{\omega} + \Sigma^{*} \times B$. Let e be the following equation with recursive variables $X = \{x, y, z\}$:

$$e(x) = (\sigma, y), \; e(y) = (\tau, x) \quad and \quad e(z) = (\sigma\tau, b).$$

The solution $e^{\dagger} : X \to \Sigma^{\omega} + \Sigma^{*} \times B$ is, clearly, given by

$$e^{\dagger}(x) = (\sigma\tau)^{\omega}, \; e^{\dagger}(y) = (\tau\sigma)^{\omega} \quad and \quad e^{\dagger}(z) = (\sigma\tau, b).$$

For a general equation $e : X \to \Sigma \times X + (\Sigma^{\omega} + \Sigma^{*} \times B)$ the solution $e^{\dagger} : X \to \Sigma^{\omega} + \Sigma^{*} \times B$ is defined by the following corecursion: if $e(x)$ lies in the right-hand summand $\Sigma^{\omega} + \Sigma^{*} \times B$ then simply $e^{\dagger}(x) = e(x)$. Otherwise we have a pair $(\sigma_1, x_1) \in \Sigma \times X$ with $e(x) = (\sigma_1, x_1)$ and $e^{\dagger}$ assigns to x the word assigned to x_1 with the prefix σ_1 added:

$$e^{\dagger}(x) = \sigma_1 e^{\dagger}(x_1).$$

   (4) Given an arbitrary polynomial functor H_{Γ} of Example 13(5) the functor $H_{\Gamma}(-) + B$ is also polynomial for the signature $\Gamma + B$, where B consists of constants. Thus ΨB is the algebra for all Γ-trees over B, i.e., leaves are labelled in the set $\Gamma_0 + B$ and nodes with $n > 0$ successors in Γ_n.

   A solution of an equation $e : X \to H_{\Gamma}X + (B + \Psi B)$ is the map $e^{\dagger} : X \to B + \Psi B$ assigning to a variable x with $e(x)$ in the right-hand smmand the value $e^{\dagger}(x) = e(x)$. For the left-hand summand we have

$$e(x) = \sigma(x_1, ..., x_n)$$

for some n-ary symbol σ, and $e^{\dagger}(x)$ is the tree with root labelled by σ and with the n maximum subtrees given by $e^{\dagger}(x_1)), ..., e^{\dagger}(x_n)$.

Remark 6. For the terminal coalgebra νF considered as an algebra we now show how solutions of equations $e : X \to FX + \nu F$ can be approximated by sequences of morphisms $e_k^{\ddagger} : X \to \mu F$ into the initial algebra. Recall that the morphisms

$\partial_k\colon \nu F \to \mu F$ of Lemma 21 converge to $\mathrm{id}_{\nu F}$. This indicates that the equation morphism e is well approximated by the following sequence of equation morphisms in μF:

$$e_k \equiv X \xrightarrow{\ e\ } FX + \nu F \xrightarrow{\ FX + \partial_k\ } FX + \mu F \quad (k < \omega). \tag{4.1}$$

We use them to define approximate solutions of e:

Definition 26. *Let an equation in νF be given, $e\colon X \to FX + \nu F$. We define its approximate solutions $e_k^{\ddagger}\colon X \to \mu F$ by induction as follows: first put*

$$e_0^{\ddagger} \equiv X \xrightarrow{\ !\ } 1 \xrightarrow{\ p\ } F0 \xrightarrow{\ w_{1,\omega}\ } \mu F. \tag{4.2}$$

Given $e_k^{\ddagger}$ define $e_{k+1}^{\ddagger}$ as the following composite:

$$\begin{array}{ccc}
X & \xrightarrow{\quad e_{k+1}^{\ddagger} \quad} & \mu F \\[2pt]
{\scriptstyle e_k}\Big\downarrow & & \Big\uparrow{\scriptstyle [\varphi,\mu F]} \\[2pt]
FX + \mu F & \xrightarrow[\ Fe_k^{\ddagger} + \mu F\ ]{} & F(\mu F) + \mu F
\end{array} \tag{4.3}$$

where $e_k = (FX + \partial_k) \cdot e$.

Observe that the first approximation $e_0^{\ddagger}$ does not use the algebra structure of μF. For the subsequent approximations the square defining solutions in Definition 24 is quite analogous to (4.4). We see that, however, $e_k^{\ddagger}$ is *not* a solution of e_k in μF in general.

Proposition 27. *The solution of every equation $e\colon X \to FX + \nu F$ in the terminal coalgebra (considered as algebra) is a limit of approximate solutions. More precisely, we have*

$$U e^{\dagger} = \lim_{k \to \infty} U(m \cdot e_k^{\ddagger}).$$

in the power of the ultrametric space $U(\nu F)$ to UX.

We are going to prove a more general result about approximate solutions for equation in ΨB, the free completely iterative algebra, below. Since for $B = 0$ we have $\nu F = \Psi 0$, the above proposition then follows.

Example 28. *Consider the equation of Example 25(3). We are given a pointed set B with $p \in B$. Here $e_0^{\ddagger}$ is constant with the value $(\varepsilon, p) \in \Sigma^* \times B$, thus, $e_1^{\ddagger}$ takes x and z to (σ, p) and y to (τ, p), using the first letter of the value of $e^{\dagger}$. Then $e_2^{\ddagger}$ takes x to $(\sigma\tau, p)$ and y to $(\tau\sigma, p)$, whereas z is taken to $(\sigma\tau, b)$. In general, $e_k^{\ddagger}$ for $k \geq 2$ takes x and y to (w, p) where w is the prefix of length k of $(\sigma\tau)^{\omega}$ and $(\tau, \sigma)^{\omega}$, resp., and it takes z to $(\sigma\tau, b)$.*

Notation 29. *The free algebra on an object B is denoted by ΦB. We show below that this is precisely the initial algebra of F_B:*

$$\Phi B = \mu X. FX + B.$$

The components of the algebra structure $F(\Phi B) + B \to \Phi B$ are denoted by $\varphi_B \colon F(\Phi B) \to \Phi B$ and $\eta'_B \colon B \to \Phi B$, resp. The first one makes ΦB an algebra for F.

The algebra ΦB is free on B w.r.t. $\eta'_B \colon B \to \Phi B$. Indeed, given an algebra $\alpha \colon FA \to A$ for F together with a morphism $f \colon B \to A$, we obtain an algebra $[\alpha, f] \colon FA + B \to A$ for F_B. And a morphism $h \colon \Phi B \to A$ is a homomorphism for F_B iff it is a homomorphism for F satisfying $h \cdot \eta'_B = f$.

We now proceed analogously to Definition 26, defining approximate solutions of equations in ΨB. This means that we move from F to F_B. Let $\partial_k^B \colon \Psi B \to \Phi B$ be the morphisms of Notation 19 related to F_B. Then every equation morphism $e \colon X \to F_B X + \Psi B$ yields morphisms

$$e_k = (F_B X + \partial_k^B) \cdot e \colon X \to FX + \Phi B$$

as in (4.1) above.

Definition 30. *Let B be a pointed object. For every equation $e \colon X \to FX + \Psi B$ in the free completely iterative algebra its approximate solutions in the free algebra ΦB are the morphisms $e_k^\ddagger \colon X \to \Phi B$ defined by induction as follows: first put*

$$e_0^\ddagger \equiv X \xrightarrow{\;!\;} 1 \xrightarrow{\;p_B\;} F_B 0 \xrightarrow{\;w_{1,\omega}\;} \mu F_B = \Phi B.$$

Given $e_k^\ddagger$ define $e_{k+1}^\ddagger$ as the following composite

$$
\begin{array}{ccc}
X & \xrightarrow{\;\;e_{k+1}^\ddagger\;\;} & \Phi B \\[2pt]
{\scriptstyle e_k}\Big\downarrow & & \Big\uparrow{\scriptstyle [\varphi_B, \Phi B]} \\[2pt]
FX + \Phi B & \xrightarrow[\;Fe_k^\ddagger + \Phi B\;]{} & F(\Phi B) + \Phi B
\end{array}
\qquad (4.4)
$$

We denote by $m_B \colon \Phi B \to \Psi B$ the morphism of Proposition 4 related to the functor F_B.

Theorem 31. *The solution of every equation $e \colon X \to FX + \Psi B$ in the free completely iterative algebra ΨB is a limit of the sequence of approximate solutions $e_k^\ddagger \colon X \to \Phi B$ in the free algebra ΦB. More precisely*

$$Ue^\dagger = \lim_{k \to \infty} U(m_B \cdot e_k^\ddagger).$$

in the power of the ultrametric space $U(\Psi B)$ to UX.

Proof. There is an ordinal i with $\Psi B = F_B^i 1$, see Theorem 3 applied to F_B, and for the connecting morphism $v_{i,k}: F_B^i 1 \to F_B^k 1 (k < \omega)$ we prove $v_{i,k} \cdot e^\dagger = v_{i,k} \cdot m_B \cdot e_k^\dagger$. Then the statement follows since $Ue^\dagger$ has distance less than 2^{-k} from $U(m_B \cdot e_k^\dagger)$. We proceed by induction, the case $k = 0$ is trivial.

From $v_{i,k} \cdot m_B \cdot e_k^\dagger = v_{i,k} \cdot e^\dagger$ we derive $v_{i,k+1} \cdot m_B \cdot e_{k+1}^\dagger = v_{i,k+1} \cdot e^\dagger$ as follows. By definition of e_k and $e_k^\dagger$ the left-hand side is

$$v_{i,k+1} \cdot m_B \cdot [\varphi_B, \Phi B] \cdot (Fe_k^\dagger + \Phi B) \cdot (FX + \partial_k^B) \cdot e.$$

Recall from (2.1) that $m_B \cdot \varphi_B = v_{i+1,i} \cdot Fm_B$. Thus,

$$v_{i,k+1} \cdot m_B \cdot \varphi_B = v_{i+1,k+1} \cdot Fm_B = F(v_{i,k} \cdot m_B).$$

The left-hand side can thus be simplified to

$$\big[F(v_{i,k} \cdot m_B), v_{i,k+1} \cdot m_B\big] \cdot (Fe_k^\dagger + \partial_k^B) \cdot e = \big[F(v_{i,k} \cdot m_B \cdot e_k^\dagger), v_{i,k+1} \cdot m_B \cdot \partial_k^B\big] \cdot e.$$

By induction hypothesis and Lemma 21 this yields

$$v_{i,k} \cdot m_B \cdot e_k^\dagger = \big[F(v_{i,k} \cdot e^\dagger), v_{i,k+1}\big] \cdot e.$$

The right-hand side yields the same result. Indeed, we have

$$v_{i,k+1} \cdot e^\dagger = v_{i,k+1} \cdot [\tau_B^{-1}, \Psi B] \cdot (Fe^\dagger + \Psi B) \cdot e$$

by definition of $e^\dagger$. Since $\tau_B^{-1} = v_{i+1,i}$ and $v_{i+1,k+1} = Fv_{i,k}$, we get

$$v_{i,k+1} \cdot e^\dagger = \big[F(v_{i,k} \cdot e^\dagger), v_{i,k+1}\big] \cdot e.$$

A completely analogous result holds for solutions in the free iterative algebra on B which we denote by RB. Recall from [7] that the difference between 'iterative' and 'completely iterative' is the restriction of the objects X of recursive variables to finitely presentable ones:

Definition 32. (see [7]). *An algebra $\alpha: FA \to A$ is called iterative if every equation morphism $e: X \to FX + A$ wit a finitely presentable object X has a unique solution.*

Example 33. (see [7]). *(1) In Example 13(3) we have seen that for deterministic automata all formal languages form the terminal coalgebra. Here ρF consists of all the regular languages.*

(2) The colimit ϱF of all coalgebras $\gamma: C \to FC$ with C finitely presentable is an iterative algebra. More precisely, denote by $\gamma^\#: C \to \varrho F$ the colimit cocone. The unique morphism $i: \varrho F \to F(\varrho F)$ making all $\gamma^\#$ coalgebra homomorphisms is invertible. And the algebra $(\varrho F, i^{-1})$ is iterative. Indeed, this is the initial iterative algebra.

(3) As a concrete example, for the polynomial functor H_Γ we have seen that the terminal coalgebra consists of all Γ-trees. ρH_Γ is the subcoalgebra of

all rational trees, which are those having up to isomorphism only finitely many subtrees. Thus e.g. for the set functor $FX = \Sigma \times X + B$ we have ϱF as the subalgebra of $\nu F = \Sigma^\omega + \Sigma^* \times B$ (see Example 13(2)) consisting of all eventually periodic words in Σ^ω (i.e., words of the form uv^ω) and all elements of $\Sigma^* \times B$.

Notation 34. *(1) The initial iterative algebra of $F_B = F(-) + B$ is denoted by $RB = \varrho F_B$. Its algebra structure $F(RB) + B \to RB$ has components denoted by $i_B\colon F(RB) \to RB$ and $\hat\eta\colon B \to RB$.*

(2) We denote by $\bar m_B\colon \Phi B \to RB$ the unique algebra homomorphism for F_B and by $s_B\colon RB \to \Psi B$ the unique coalgebra homomorphism. Since ΨB is a terminal coalgebra, the homomorphism $m_B\colon \Phi B \to \Psi B$ is their composite:

$$m_B = s_B \cdot \bar m_B.$$

Since m_B is monic by Proposition 4, we see that $\bar m_B$ is monic.

For the functor H_Γ of Example 13(5) $s_B\colon RB \to \Psi B$ is the inclusion of the set of all rational Γ-trees on B.

Proposition 35 ([7], **3.9).** *The algebra $i_B\colon F(RB) \to RB$ is a free iterative algebra for F w.r.t. $\hat\eta\colon B \to RB$. If finitely generated objects of $\mathcal{K}$ coincide with finitely presentable ones, then $s_B\colon RB \to \Psi B$ is a monomorphism.*

Thus, we have subobjects $\Phi B \overset{\bar m_B}{\hookrightarrow} RB \overset{s_B}{\hookrightarrow} \Psi B$ and $U(RB)$ is a canonical ultrametric space, as a subspace of $U(\Psi B)$.

Definition 36. *Let $e\colon X \to FX + RB$ be an equation in the free iterative algebra with X finitely presentable. We define equations e_k in the free algebra ΦB as follows:*

$$e_k \equiv X \xrightarrow{\ e\ } FX + RB \xrightarrow{\ FX + s_B\ } FX + \Psi B \xrightarrow{\ FX + \partial_k\ } FX + \Phi B$$

The approximate solutions $e_k^\dagger\colon X \to \mu F$ of e are then defined as in Definition 26.

Corollary 37. *Suppose that finitely generated and finitely presentable objects coincide. For every equation $e\colon X \to FX + RB$ in the free iterative algebra (X finitely presentable), its solution is a limit of approximate solutions in the free algebra. More precisely*

$$Ue^\dagger = \lim_{k \to \infty} U(\bar m_B \cdot e_k^\dagger).$$

in the power of the ultrametric space $U(RB)$ to UX.

This follows from Theorem 31 applied to the equation

$$\bar e \equiv X \xrightarrow{\ e\ } FX + RB \xrightarrow{\ FX + s_B\ } FX + \Psi B.$$

Indeed, on the one hand, we have $\bar e_k^\dagger = e_k^\dagger$ for every k (by an easy induction). We verify below that $\bar e^\dagger = s_B \cdot e^\dagger$. Then the corollary follows from $m_B = s_B \cdot \bar m_B$ since s_B is monic.

Thus we are to prove that $s_B \cdot e^\dagger$ solves $\bar{e}$. This is clear from the diagram below

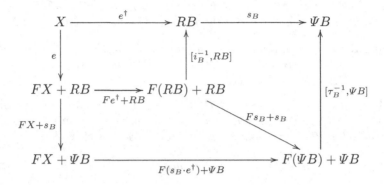

Indeed, the left-hand square commutes since $e^\dagger$ solves e, the righ-hand part commutes since $s_B \colon (RB, i_B^{-1}) \to (\psi B, \tau_B^{-1})$ is a homomorphism of algebras for F_B. The lower part clearly commutes. Thus, the outward square commutes, and its left-hand vertical arrow is $\bar{e}$.

5   Conclusions

We have studied colagebras for finitary endofunctors preserving monomorphisms on a wide collection of categories such as sets, vector spaces, posets, boolean algebras etc. Each such endofunctor F carries a canonical ultrametric on the underlying set of the terminal coalgebra νF. Moreover, if F is grounded, i.e., $F0$ has a global element, the underlying set of the initial algebra μF is a dense subspace of that ultrametric space. Given a coalgebra on an object A, we have presented a canonical Cauchy sequence of morphisms from A to μF which converges to the unique coalgebra homomorphism from A to νF in the power of the above ultrametric space νF.

For every pointed object B we formed the free iterative algebra RB and the free completely iterative algebra ΨB on B. We derived a canonical ultrametric on the underlying set of each of them. And the free F-algebra ΦB forms a dense subspace of both RB and ΨB. Moreover, for every recursive equation e in ΨB we have presented a canonical Cauchy sequence of morphisms into ΦB converging to the unique solution $e^\dagger$ of e in ΨB (again, in the corresponding power of the space $U(\Psi B)$). Analogously for recursive equations with finitely presentable variable-objects in the algebra RB.

References

1. Adámek, J.: Free algebras and automata realizations in the language of categories. Comment. Math. Univ. Carolinae **15**, 589–602 (1974)
2. Adámek, J.: Final coalgebras are ideal completions of initial algebras. J. Logic Comput. **2**, 217–242 (2002)

3. Adámek, J.: On final coalgebras of continuous functors. Theor. Comput. Sci. **294**, 3–29 (2003)
4. Adámek, J.: On terminal coalgebras derived from initial algebras. In: proceedings of the Coalgebraic and Algebraic Methods in Computer Science (CALCO 2019), LIPIcs, vol. 139 (2019)
5. Adámek, J.: On free completely iterative algebras. In: proceedings of the Computer Science Logic (CSL 2020), LIPIcs, vol. 152 (2020)
6. Adámek, J., Milius, S., Sousa, L., Wißmann, T.: On finitary functors. Theory Appl. Categ. **34**, 1134–1164 (2019)
7. Adámek, J., Millius, S., Velebil, J.: Iterative algebras at work. Math. Struct. Comput. Sci. **16**, 1085–1131 (2006)
8. Adámek, J., Rosický, J.: Locally Presentable and Accessible Categories. Cambridge University Press, Cambridge (1994)
9. Adámek, J., Trnková, V.: Relatively terminal coalgebras. J. Pure Appl. Algebra **216**, 1887–1895 (2012)
10. Barr, M.: Terminal coalgebras in well-founded set theory. Theoret. Comput. Sci. **114**, 299–315 (1993)
11. Milius, S.: Completely iterative algebras and completely iterative monads. Inf. Comput. **196**, 1–41 (2005)
12. Rutten, J.: Universal coalgebra. Theor. Comput. Sci. **249**, 3–80 (2000)
13. Worrell, J.: On the final sequence of a finitary set functor. Theor. Comput. Sci. **338**, 184–199 (2005)

Duality for Instantial Neighbourhood Logic via Coalgebra

Nick Bezhanishvili[1], Sebastian Enqvist[2], and Jim de Groot[3]([envelope]) [ORCID]

[1] University of Amsterdam, Amsterdam, The Netherlands
n.bezhanishvili@uva.nl
[2] Stockholm University, Stockholm, Sweden
sebastian.enqvist@philosophy.su.se
[3] The Australian National University, Canberra, Australia
jim.degroot@anu.edu.au

Abstract. Instantial Neighbourhood Logic (INL) has been introduced recently as a language for neighbourhood frames where existential information can be given about what kind of worlds occur in a neighbourhood of a current world. Apart from its semantics, its proof theory and bisimulation games have also been studied. However, conspicuously absent from the treatment of INL is the notion of *descriptive* frames.

This is the gap that we are closing in this paper. We introduce descriptive frames for INL and we prove that these are dual to boolean algebras with instantial operators (BAIOs), which give the algebraic semantics of INL. Our methods for establishing this duality make essential use of coalgebra: we observe that BAIOs are algebras for a functor on the category of boolean algebras and show that this functor is dual to the double Vietoris functor (i.e. the composition of the Vietoris functor with itself), thus obtaining a dual equivalence between double Vietoris coalgebras and BAIOs. The proof of our main result is then completed by showing that double Vietoris coalgebras correspond precisely to descriptive frames. As a corollary we obtain that every extension of INL is sound and complete with respect to descriptive frames, that descriptive frames enjoy the Hennessy-Milner property, and as a result, that finite neighbourhood frames enjoy the Hennessy-Milner property.

Keywords: Duality · Modal logic · Instantial neighbourhood logic · Descriptive frames · Coalgebra

1 Introduction

The concept of *duality* arises in many areas of mathematics, logic and computer science. The first duality milestone in algebraic logic was the Stone representation theorem [31], which described the categorical duality between boolean algebras and so-called Stone spaces. Subsequently, many representation and duality theorems have been established, including a representation theorem for Riesz

© IFIP International Federation for Information Processing 2020
Published by Springer Nature Switzerland AG 2020
D. Petrişan and J. Rot (Eds.): CMCS 2020, LNCS 12094, pp. 32–54, 2020.
https://doi.org/10.1007/978-3-030-57201-3_3

spaces [38], Priestley duality for distributive lattices [28], and Esakia duality for (bi-)Heyting algebras [9,12–14].

The first representation theorem in the realm of modal logic was given by Jónsson and Tarski [20]. Although *op. cit.* does not mention modal logic explicitly, it introduces relational semantics and uses the representation theorem to relate algebraic and relational semantics. The full duality between boolean algebras with operators and descriptive general frames was established by Goldblatt [16], and many other dualities ensued [8,15,19,25].

These dualities are useful because they provide two, algebraic and geometric, perspectives on the same object and allow one to translate results from the algebraic language to the geometric one via (descriptive) frames, and vice versa. For example, this was a key ingredient in Sambin and Vaccaro's simplified proof of the celebrated Sahlqvist completeness theorem in modal logic [30].

A novel system of modal logic for reasoning about neighbourhood frames, called *Instantial Neighbourhood Logic* (INL), was recently introduced in [3]. The modalities in INL give existential information about what kind of worlds occur in a neighbourhood of a current world. Motivations for investigating this logic range from topology to games, and from modelling notions of evidence to belief revision. Surprisingly however, a duality result is still absent from the theory of INL, despite an abundance of recent interest in the logic [1,2,32,39,40].

This is the gap we are filling in this paper. We introduce a suitable notion of descriptive frames for INL and prove a (categorical) duality between descriptive INL-frames and boolean algebras with instantial operators (BAIOs). The latter give the algebraic semantics for INL and play the same rôle for INL that boolean algebras with operators play for normal modal logic [7,10] and boolean algebras with monotone operators for monotone modal logic [17,18].

Our main technical tool for establishing this duality is *coalgebra*. Coalgebras arise as the dual notion of *algebras for a functor* and have found many applications in logic and computer science. For example, they are a natural setting for dealing with non-wellfounded data structures such as streams or infinite trees, and, triggered by Moss' paper on coalgebraic logic [26], they have become a widely-used framework for describing semantics for modal logics [21,22,24,29].

A key feature of the coalgebraic perspective on duality is that, in a sense, it isolates the essential part of the duality. In the case of modal logic, the traditional approach of Jónsson-Tarski duality considers modal algebras as boolean algebras with operators, and directly constructs the dual Kripke frames. From the point of view of coalgebra, this turns out to really be a duality between *functors*. Once we identify modal algebras as the algebras for a certain functor on the category of boolean algebras and recognise its dual functor as the Vietoris functor, the duality between algebras and coalgebras for these two functors, which piggy-backs Stone duality, is a trivial consequence. The same approach works well for other variations of modal logic such as monotone modal logic [18] or positive modal logic (which piggy-backs on Priestley duality) [27]. While these examples have been re-cast in this style with hindsight, we believe that this coalgebraic approach serves as a useful blueprint for establishing dualities for new logics.

If a logical system and its algebras are given, and a natural candidate for dual "descriptive frames" is suggested, then proceed as follows:

1. Describe the algebras of the logic as algebras for a functor.
2. Find the dual functor T.
3. Show that T-coalgebras are equivalent to descriptive frames.

In the current paper we use this strategy to prove a duality theorem for INL. First, we observe that BAIOs are algebras for a functor on the category of boolean algebras and homomorphisms. Second, we recall that it was observed in [3] that the neighborhood semantics of INL corresponds to coalgebras for the *double covariant powerset functor*. Since Kripke frames are coalgebras for the covariant powerset functor, and modal algebras are dual to Vietoris coalgebras, this suggests that BAIOs should be dual to coalgebras for the *double Vietoris functor*. We verify that this is indeed the case. Lastly, we show that our descriptive frames are precisely the "double Vietoris coalgebras", in the sense that the categories are isomorphic. Hence, these descriptive frames are dual to BAIOs.

A rather trivial step in this proof is the representation of BAIOs as algebras for a functor I; the functor is simply read off from the defining equations of BAIOs. It was observed by Lutz Schröder (private communication) that the logic INL can be translated into the composition of standard modal logic with itself, and vice versa, in a semantics-preserving manner. This suggests that we could optionally have represented BAIOs as algebras for the functor M∘M, where M is the functor whose algebras correspond to modal algebras. Since M is dual to the Vietoris functor, the dual equivalence of M ∘ M-algebras with double Vietoris coalgebras would follow directly (in fact, as a special case of a more general result in [23]). While this alternative approach has a certain elegance to it, we have opted for a more direct representation of INL-algebras here. The main benefit is that our proof gives a quite standard canonical model construction for INL and its extensions, without a detour via a translation. This makes the duality better suited for dealing with issues like canonicity and Sahlqvist completeness, which we hope to address in future work. We obtain the equivalence of INL-algebras with algebras for M ∘ M as a corollary. Our proof also gives a representation of the double Vietoris functor where the topology is described in terms of the INL-modalities, rather than two layers of box and diamond modalities.

As a corollary of our main duality theorem we obtain that every extension of INL is sound and complete with respect to descriptive frames, that descriptive frames enjoy the Hennessy-Milner property, and as a result, that finite INL-frames enjoy the Hennessy-Milner property.

Outline. The paper is structured as follows: In Sect. 2 we recall the definitions of Instantial Neighbourhood Logic and its semantics, algebras and coalgebras, Stone duality, and the Vietoris functor. In Sect. 3 we define boolean algebras with instantial operators and descriptive INL-frames. The main results of the paper appear in Sect. 4, where we prove that the category of BAIOs is dually equivalent to the category of coalgebras for the double Vietoris functor, and in Sect. 5, where we identify coalgebras for the double Vietoris functor with descriptive INL-frames. Finally, we provide some applications of the developed theory in Sect. 6, and discuss possible directions for future work in Sect. 7.

2 Preliminaries

2.1 Instantial Neighbourhood Logic

We briefly recall the language and semantics of Instantial Neighbourhood Logic from [3]. The language $\mathcal{L}(\mathrm{Prop})$ of Instantial Neighbourhood Logic over some arbitrary but fixed set Prop of proposition letters is defined recursively by

$$\varphi ::= \top \mid p \mid \neg\varphi \mid \varphi \wedge \varphi \mid \Box(\varphi_1, \ldots, \varphi_n; \varphi),$$

where $p \in \mathrm{Prop}$ and $n \in \omega$. Observe that we have a countably infinite number of modal operators: one for each $n \in \omega$. Formulas in $\mathcal{L}(\mathrm{Prop})$ can be interpreted in neighbourhood frames, which we shall call *INL-frames*.

Definition 2.1. An *INL-frame* is a pair (X, N) comprised of a set X and a neighbourhood function $N : X \to \mathsf{PP}X$, where P is the (covariant) powerset functor. A *neighbourhood model* is a tuple (X, N, V) where (X, N) is a neighbourhood frame and $V : \mathrm{Prop} \to \mathsf{P}X$ is a valuation of the proposition letters.

An *INL-morphism* from (X, N) to (X', N') is a map $f : X \to X'$ such that

$$N'(f(x)) = \{f[a] \mid a \in N(x)\},$$

for all $x \in X$. Here $f[a]$ denotes the direct image of a under f. An *INL-morphisms* $(X, N, V) \to (X', N', V')$ between neighbourhood models is an INL-morphism f between the underlying frames which additionally satisfies

$$x \in V(p) \quad \text{iff} \quad f(x) \in V'(p)$$

for every $p \in \mathrm{Prop}$.

Write **INL** and **INL**^{M} for the categories of neighbourhood frames and neighbourhood models, respectively, with their corresponding notion of morphism.

The interpretation of INL-formulas in a neighbourhood model (X, N, V) is defined recursively, where the classical connectives are treated in the standard manner. For the modalities, let $x \Vdash \Box(\varphi_1, \ldots, \varphi_n; \psi)$ if there is a neighbourhood $w \in N(x)$ of x such that $y \Vdash \psi$ for all $y \in w$, and for each φ_i there is $y_i \in w$ such that $y_i \Vdash \varphi_i$. We write $[\![\varphi]\!] = \{x \in X \mid x \Vdash \varphi\}$ and say that x *satisfies* φ if $x \in [\![\varphi]\!]$. Two states are called *logically equivalent* if they satisfy precisely the same formulas.

The interpretation of the modalities can be conveniently reformulated using the following notion of *witnesses*.

Definition 2.2. Let X be a set and $w, a_1, \ldots, a_n, b$ subsets of X. We say that w *witnesses* $(a_1, \ldots, a_n; b)$ if and only if $w \cap a_i \neq \emptyset$ for all $i \in \{1, \ldots, n\}$ and $w \subseteq b$. We say that w *co-witnesses* $(a_1, \ldots, a_n; b)$ if and only if $w \subseteq a_i$ for some $i \in \{1, \ldots, n\}$ or $w \cap b \neq \emptyset$.

It is straightforward to see that in a neighbourhood model (X, N, V) a state x satisfies $\Box(\varphi_1, \ldots, \varphi_n; \psi)$ if there is a $w \in N(x)$ witnessing $(\llbracket \varphi_1 \rrbracket, \ldots, \llbracket \varphi_n \rrbracket; \llbracket \psi \rrbracket)$.

Definition 2.3. Let (X, N) be a neighbourhood frame. Define the map $m_{\Box,n}$: $(\mathsf{P}X)^{n+1} \to \mathsf{P}X$ by

$$m_{\Box,n}(a_1, \ldots, a_n; b) = \{x \in X \mid \exists w \in N(x) \text{ which witnesses } (a_1, \ldots, a_n; b)\}.$$

When there is no danger of confusion, we suppress the subscript n from $m_{\Box,n}$.

Yet another way to view the interpretation of the modalities in a neighbourhood model is via the equality $\llbracket \Box(\varphi_1, \ldots, \varphi_n; \psi) \rrbracket = m_\Box(\llbracket \varphi_1 \rrbracket, \ldots, \llbracket \varphi_n \rrbracket, \llbracket \psi \rrbracket)$.

2.2 Stone Duality

Write **BA** for the category of boolean algebras and homomorphisms, and **Stone** for the category of Stone spaces and continuous functions.

The contravariant functor uf : **BA** $\to$ **Stone** takes a boolean algebra B to the collection uf$B$ of ultrafilters of B topologized by the base $\tilde{B} = \{(\!|b|\!) \mid b \in B\}$, where $(\!|b|\!) = \{u \in \mathsf{uf}B \mid b \in u\}$. The action of uf on a homomorphism $h : B \to B'$ in **BA** is defined by $(\mathsf{uf}h)(u') = h^{-1}(u')$. In the converse direction, the contravariant functor clp : **Stone** $\to$ **BA** takes a Stone space to its boolean algebra of clopens and a continuous function to its inverse. The functors uf and clp constitute a dual equivalence between **BA** and **Stone**.

2.3 Algebra, Coalgebra, and the Vietoris Functor

We recall the definitions of algebras, coalgebras, and the Vietoris functor.

Definition 2.4. Let F be an endofunctor on a category **C**. An F-*coalgebra* is a pair (c, γ) such that $\gamma : c \to \mathsf{F}c$ is a morphism in **C**. An F-*coalgebra morphism* $(c, \gamma) \to (c', \gamma')$ is a morphism $f : c \to c'$ in **C** satisfying $\gamma' \circ f = \mathsf{F}f \circ \gamma$. The collection of F-coalgebras and F-coalgebra morphisms constitutes the category **Coalg**(F).

The dual notion of a coalgebra is that of an algebra: an F-algebra is a morphism $\gamma : \mathsf{F}c \to c$ in **C**, an F-algebra morphism $(c, \gamma) \to (c', \gamma')$ is a morphism $f : c \to c'$ in **C** such that $\gamma' \circ \mathsf{F}f = f \circ \gamma$, and they form the category **Alg**(F).

Coalgebras for an endofunctor on **Set** are used to describe *systems* [29], but can also be used to characterise the frame semantics for a wide variety of modal logics [24]. In particular, as noted in [3, Section 7.5], we have:

Proposition 2.5. *The category* **INL** *of INL-frames and INL-morphisms is isomorphic to* **Coalg**(PP).

In the realm of modal logic a well-known example of a category of algebras is the category **MA** of modal algebras: we have **MA** $\cong$ **Alg**(M), where M is an endofunctor on the category **BA** of boolean algebras and homomorphisms defined as follows:

Definition 2.6. For a boolean algebra B let $\mathsf{M}B$ be the free boolean algebra generated by the set $\{\Box a \mid a \in B\}$, modulo the relations $\Box a \wedge \Box b = \Box(a \wedge b)$ and $\Box\top = \top$. This assignment extends to an endofunctor on **BA** by defining the action of M on a homomorphism $h : B \to B'$ via $\mathsf{M}h(\Box a) = \Box h(a)$ and extending this (uniquely) to $\mathsf{M}B$.

A prime example of a category of coalgebras for an endofunctor on a category different from **Set** is the category **DGF** of descriptive general frames and appropriate morphisms. This is isomorphic to the category of coalgebras for the *Vietoris functor* V on **Stone**, the category of Stone spaces and continuous functions [23]. We recall the definition of the Vietoris functor on **Top** (originally introduced by Leopold Vietoris for compact Hausdorff spaces [37], see also [19, Section III.4] for a localic perspective), which restricts to the Vietoris functor V on **Stone**.

Definition 2.7. For a topological space $\mathbb{X}$ let $\mathsf{V}'\mathbb{X}$ be the set of compact subsets of $\mathbb{X}$ topologized by

$$\boxdot a = \{b \in \mathsf{V}'\mathbb{X} \mid b \subseteq a\}, \quad \Diamond a = \{b \in \mathsf{V}'\mathbb{X} \mid b \cap a \neq \emptyset\},$$

where a ranges over the open sets of $\mathbb{X}$. This is called the *Vietoris topology*. For a continuous function $f : \mathbb{X} \to \mathbb{X}'$ define $\mathsf{V}'f : \mathsf{V}'\mathbb{X} \to \mathsf{V}'\mathbb{X}' : c \mapsto f[c]$, i.e. $\mathsf{V}'f$ is the direct image of f. Then V' defines a functor on **Top** called the *Vietoris functor*.

It is well known that V' restricts to an endofunctor on **Stone**, which we denote by V. Moreover, if $\mathbb{X}$ is a Stone space then the topology on $\mathsf{V}\mathbb{X}$ is generated by $\boxdot a, \Diamond a$, where a ranges over the *clopen* subsets of $\mathbb{X}$. The fact that V is the (Stone) dual of M then implies

$$\mathbf{MA} \cong \mathbf{Alg}(\mathsf{M}) \equiv^{\mathrm{op}} \mathbf{Coalg}(\mathsf{V}) \cong \mathbf{DGF},$$

where we use $\cong$ to indicate an isomorphism of categories and $\equiv^{\mathrm{op}}$ for a dual equivalence. For details we refer to [23].

3   BAIOs and General Frames

In this section we define boolean algebras with instantial operators (BAIOs) and general INL-frames.

A boolean algebra with instantial operators comprises of a boolean algebra B and an ω-indexed family of functions, reflecting the infinite number of modal operators in INL, and provide algebraic semantics for instantial neighbourhood logic. They play the same rôle for INL that modal algebras play for normal modal logic [7,10] and boolean algebras with monotone operators for monotone modal logic [17,18].

A general INL-frame is an INL-frame together with a collection of "admissible subsets" which is closed under certain operations. As usual, these admissible subsets form a BAIO. In the converse direction, we will show that every BAIO gives rise to a general INL-frame.

Definition 3.1. A *boolean algebra with instantial operators* is a pair $(B, (f_n)_{n \in \omega})$ consisting of a boolean algebra B and an ω-indexed set of functions $f_n : B^{n+1} \to B$ satisfying the following equations for all $n \in \omega$:

(B_1) $f_n(a_1, \ldots, a_{n-1}, \bot; b) = \bot$;

(B_2) $f_n(a_1, \ldots, a_i, a_{i+1}, \ldots, a_n; b) = f_n(a_1, \ldots, a_{i+1}, a_i, \ldots, a_n; b)$;

(B_3) $f_n(a_1, \ldots, a_n; b) \leq f_n(a_1, \ldots, a_n \vee a_n'; b \vee b')$;

(B_4) $f_n(a_1, \ldots, a_n; b) \leq f_n(a_1, \ldots, a_n \wedge b; b)$;

(B_5) $f_n(a_1, \ldots, a_n; b) \leq f_{n+1}(a_1, \ldots, a_n, c; b) \vee f_n(a_1, \ldots, a_n; b \wedge \neg c)$;

(B_6) $f_{n+1}(a_1, \ldots, a_{n+1}; b) \leq f_n(a_1, \ldots, a_n; b)$;

(B_7) $f_n(a_1, \ldots, a_n; b) \leq f_{n+1}(a_1, \ldots, a_n, a_n; b)$.

A morphism between BAIOs $(B, (f_n)_{n \in \omega})$ and $(B', (f_n')_{n \in \omega})$ is a boolean algebra homomorphism $h : B \to B'$ which satisfies

$$h(f_n(a_1, \ldots, a_n; b)) = f_n'(ha_1, \ldots, ha_n; hb)$$

for all $a_i, b \in B$ and $n \in \omega$. The collection of BAIOs and BAIO morphisms forms a category (a variety of algebras, in fact) denoted by **BAIO**.

Every INL-frame (X, N) gives rise to a BAIO, namely its complex algebra.

Example 3.2. Let (X, N) be an INL-frame. Let $\mathsf{P}X$ be the powerset of X viewed as a boolean algebra and define $f_n(a_1, \ldots, a_n; b) = m_\square(a_1, \ldots, a_n; b)$. Then it is easy to verify that $(\mathsf{P}X, (f_n)_{n \in \omega})$ is a BAIO. This is called the *complex algebra* of (X, N).

Example 3.3. Recall that Prop is an arbitrary but fixed set of proposition letters and let $\mathcal{L} = \mathcal{L}(\text{Prop})$ be the collection of instantial formulas as defined in Subsect. 2.1. Write $\varphi \equiv \psi$ if two formulas are provably equivalent in the axiomatization given in [3, Section 4] and write $[\varphi]$ for the equivalence class of φ under $\equiv$. Then $\mathcal{L}/\equiv$ is a BAIO, where $f_n([\varphi_1], \ldots, [\varphi_n]; [\psi])$ is defined to be $[\square(\varphi_1, \ldots, \varphi_n; \psi)]$. This is of course the free BAIO generated by Prop, and is known as the Lindenbaum-Tarski algebra.

Towards a duality theorem for BAIOs, we define general INL-frames. These are INL-frames together with a subalgebra of their complex algebra.

Definition 3.4. A *general INL-frame* is a triple (X, N, A) such that (X, N) is an INL-frame and $A \subseteq \mathsf{P}X$ is a collection of admissible sets that is closed under boolean operations and the operation $m_\square : (\mathsf{P}X)^{n+1} \to \mathsf{P}X$ (see Definition 2.3).

A *general INL-morphism* from (X, N, A) to (X', N', A') is an INL-morphism $f : (X, N) \to (X', N')$ satisfying $f^{-1}(a') \in A$ for all $a' \in A'$. Write **G-INL** for the category of general INL-frames and general INL-morphisms.

Since every algebra is a subalgebra of itself, every INL-frame can be seen as a general INL-frame:

Example 3.5. If (X, N) is an INL-frame, then setting $A = \mathsf{P}X$ yields a general INL-frame.

Example 3.6. For a BAIO $(B, (f_n)_{n \in \omega})$, let $\mathsf{uf}B$ be the collection of ultrafilters of B and let $\widetilde{B} = \{(\!|a|\!) \mid a \in B\}$, where $(\!|a|\!) = \{u \in \mathsf{uf}B \mid a \in u\}$. Define a neighbourhood function N for $\mathsf{uf}B$ via

$$N(u) = \{d \subseteq \mathsf{uf}B \mid f_n(a_1, \ldots, a_n; b) \in u \text{ whenever}$$
$$d \text{ witnesses } ((\!|a_1|\!), \ldots, (\!|a_n|\!); (\!|b|\!))\}.$$

Then $(\mathsf{uf}B, N, \widetilde{B})$ is a general INL-frame.

In the converse direction of Example 3.6 we have the functor $\mathsf{F} : \mathbf{G\text{-}INL} \to \mathbf{BAIO}$ which sends a general INL-frame (X, N, A) to $(A, (m_{\square,n})_{n \in \omega})$.

One can now ask whether $\mathbf{G\text{-}INL}$ can be restricted to a category of *descriptive* INL-frames such that the restriction of F to these descriptive frames gives rise to a dual equivalence with $\mathbf{BAIO}$. This turns out to be the case for the following definition of descriptive INL-frames:

Definition 3.7. A general INL-frame (X, N, A) is called:

- *differentiated* if for any two distinct points $x, y \in X$ there is $a \in A$ such that $x \in a$ and $y \notin a$;
- *compact* if $\bigcap A' \neq \emptyset$ for any subset A' of A with the finite intersection property;
- *crowded* if for all $x \in X$ and $d \subseteq X$ such that $d \notin N(x)$ we can find $a_1, \ldots, a_n, b$ such that d witnesses $(a_1, \ldots, a_n; b)$ while no $d' \in N(x)$ witnesses $(a_1, \ldots, a_n; b)$.

A *descriptive INL-frame* is a general INL-frame that is differentiated, compact and crowded. Denote by $\mathbf{D\text{-}INL}$ the full subcategory of $\mathbf{G\text{-}INL}$ whose objects are descriptive INL-frames.

The notion of crowdedness is the INL analogue of the tightness condition for normal modal logic [7, Definition 5.65]. Intuitively, it states that $N(x)$ is, in a sense, determined by the admissible subsets. In passing, we make the following observation, the proof of which is straightforward.

Proposition 3.8. *Let (X, N) be an INL-frame and suppose X is finite. Then $A = \mathsf{P}X$ is the unique set of admissible sets making (X, N, A) a descriptive INL-frame.*

We aim to prove the following duality result:

Theorem 3.9. *We have a dual equivalence*

$$\mathbf{D\text{-}INL} \equiv^{\mathrm{op}} \mathbf{BAIO}.$$

In order to prove this, we adhere to the strategy suggested in the introduction. First we identify $\mathbf{BAIO}$ with the category of algebras for some functor I on $\mathbf{BA}$, then we determine the (Stone) dual of I, and finally, we show that descriptive INL-frames are precisely coalgebras for this dual functor. In a diagram:

$$
\begin{array}{ccc}
\textbf{BAIO} & \xleftarrow{\quad\text{Theorem 3.9}\quad} & \textbf{D-INL} \\
\Big\| \;\text{Theorem 4.2} & & \Big\| \;\text{Theorem 5.1} \\
\textbf{Alg}(I) & \xrightleftharpoons[\text{Theorem 4.12}]{} & \textbf{Coalg}(VV)
\end{array}
\tag{1}
$$

Concretely, the dual equivalence from Theorem 3.9 will be given by the construction from Example 3.6 and the subsequent paragraph, see Remark 5.7 below.

4 Duality

We show that BAIOs are algebras for the functor $I : \textbf{BA} \to \textbf{BA}$, and that I is the (Stone) dual of the double Vietoris functor VV on **Stone**. As a consequence we obtain an algebra/coalgebra duality in Theorem 4.12.

Definition 4.1. Let B be a boolean algebra. Abbreviate $(\boldsymbol{a}; b) = (a_1, \ldots, a_n; b)$ for an $(n+1)$-tuple of elements of B. Let IB be the boolean algebra generated by $\square(a_1, \ldots, a_n; b)$, where $n \in \omega$ and $a_i, b \in B$, subject to the relations

(I_1) $\square(\boldsymbol{a}, \bot; b) = \bot$;

(I_2) $\square(a_1, \ldots, a_i, a_{i+1}, \ldots, a_n; b) = \square(a_1, \ldots, a_{i+1}, a_i, \ldots, a_n; b)$;

(I_3) $\square(\boldsymbol{a}, c; b) \leq \square(\boldsymbol{a}, c \vee c', b \vee b')$;

(I_4) $\square(\boldsymbol{a}, c; b) \leq \square(\boldsymbol{a}, c \wedge b; b)$;

(I_5) $\square(\boldsymbol{a}; b) \leq \square(\boldsymbol{a}, c; b) \vee \square(\boldsymbol{a}; b \wedge \neg c)$;

(I_6) $\square(\boldsymbol{a}, c; b) \leq \square(\boldsymbol{a}; b)$;

(I_7) $\square(a_1, \ldots, a_n; b) \leq \square(a_1, \ldots, a_n, a_n; b)$.

For a homomorphism $f : B \to B'$ define $If : IB \to IB'$ on generators by

$$
(If)(\square(a_1, \ldots, a_n; b)) = \square(f(a_1), \ldots, f(a_n); f(b)).
$$

The assignment I determines an endofunctor on **BA**, called the *instantial functor*.

Item I_2 of Definition 4.1 allows us to put the a_i in any desired order. In I_4 equality holds because of I_3 and in I_7 equality hold because of I_6. The proof of the following theorem is standard.

Theorem 4.2. BAIO = Alg(I).

As stated in the introduction, it seems reasonable to expect that the double Vietoris functor VV is the dual of I under the dual equivalence between boolean algebras and Stone spaces. We prove that this is indeed the case. More concretely, we give a natural isomorphism

$$
\text{uf} \circ I \circ \text{clp} \to VV.
\tag{2}
$$

We first work towards an isomorphism uf $\circ$ l $\circ$ clpX $\cong$ VVX, where X is a Stone space. Lemma 4.11 then states that the collection of these isomorphisms is in fact natural. This ultimately proves the duality between (the categories of) l-algebras and VV-coalgebras (Theorem 4.12).

We commence by giving an alternative subbase for the double Vietoris topology, which is tailored to our specific needs.

Proposition 4.3. *Let* X *be a Stone space. The topology on* VVX *is generated by the clopen subbase*

$$\boxdot(a_1,\ldots,a_n;b) = \{W \in VVX \mid \exists w \in W \text{ s.t. } w \text{ witnesses } (a_1,\ldots,a_n;b)\},$$
$$\Diamond\!\!\!\Diamond(a_1,\ldots,a_n;b) = \{W \in VVX \mid \forall w \in W \text{ s.t. } w \text{ co-witnesses } (a_1,\ldots,a_n;b)\},$$

where the a_i, b *range over the clopen subsets of* X.

Proof. The given sets are clopen in VVX, because

$$\boxdot(a_1,\ldots,a_n;b) = \Diamond(\Diamond a_1 \cap \cdots \cap \Diamond a_n \cap \boxdot b)$$

and

$$\Diamond\!\!\!\Diamond(a_1,\ldots,a_n;b) = \boxdot(\boxdot a_1 \cup \cdots \cup \boxdot a_n \cup \Diamond b).$$

In order to show that they are a subbase for the topology on VVX, we must show that $\boxdot A$ and $\Diamond A$ are boolean combinations of clopens of the form $\boxdot(\boldsymbol{a};b)$ and $\Diamond\!\!\!\Diamond(\boldsymbol{a};b)$, where A is clopen in VX. Note that A can be written as the finite intersection of finite unions of clopens in VX of the form $\boxdot a, \Diamond a$. Moreover, we may assume that there is a single diamond in each finite union because diamonds distribute over unions. So we may write $A = \bigcap_{i=1}^{n}(\boxdot a_1 \cup \cdots \cup \boxdot a_{m_i} \cup \Diamond b_i)$. This implies

$$\boxdot A = \bigcap_{i=1}^{n} \boxdot\left(\boxdot a_1 \cup \cdots \cup \boxdot a_{m_i} \cup \Diamond b_i\right) = \bigcap_{i=1}^{n} \Diamond\!\!\!\Diamond(a_1,\ldots,a_{m_i};b_i).$$

Similarly, writing A as a finite union of finite intersections, $\Diamond A$ can be expressed as a finite union of clopens of the form $\boxdot(a_1,\ldots,a_n;b)$. □

We note that the statement of Proposition 4.3 holds for any topological space X if we require the a_i and b to be range over the *open* subsets of X. The proof of this is slightly more involved because it crucially uses compactness of the elements of VX.

Remark 4.4. Recall that ultrafilters correspond bijectively to homomorphisms into 2, the two-element boolean algebra. For an ultrafilter u in B let $p : B \to 2$ be the corresponding homomorphism. Then $p(b) = \top$ iff $b \in u$. We will use these two perspectives interchangeably.

We now define a map $\xi : \mathsf{uf} \circ \mathsf{l} \circ \mathsf{clpX} \to \mathsf{VVX}$, where $\mathbb{X}$ is a Stone space. These will form the components of the intended natural transformation from (2). Although such a map can be defined for any Stone space $\mathbb{X}$, and therefore the definition actually yields a **Stone**-indexed collection of maps $(\xi_\mathbb{X})_{\mathbb{X} \in \mathbf{Stone}}$, we shall refrain from writing this subscript until Theorem 4.11, where we prove that this collection is a natural transformation.

Intuitively, to an ultrafilter p we want to attach a closed subset W_p of VX that satisfies $W_p \in \boxdot(a_1, \ldots, a_n; b)$ if and only if $p(\Box(a_1, \ldots, a_n; b)) = \top$. In our definition, we guarantee the implication from left to right by "killing the witnesses": If $p(\Box(a_1, \ldots, a_n; b)) = \bot$, then we make sure that none of the witnesses of $(a_1, \ldots, a_n; b)$ is in W_p. In other words, we stipulate that W_p be disjoint from $\Diamond a_1 \cap \cdots \cap \Diamond a_n \cap \boxdot b$.

Definition 4.5. For a Stone space $\mathbb{X}$, define $\xi : \mathsf{uf} \circ \mathsf{l} \circ \mathsf{clpX} \to \mathsf{VVX}$ by sending an ultrafilter p to

$$W_p = \mathsf{VX} \setminus \bigcup \{ \Diamond a_1 \cap \cdots \cap \Diamond a_n \cap \boxdot b \mid p(\Box(a_1, \ldots, a_n; b)) = \bot \}.$$

Since W_p is the complement of a union of clopen subsets of VX, it is closed in VX, hence an element of VVX. Therefore ξ is well defined. For the converse direction we need the following definition.

Definition 4.6. For a Stone space $\mathbb{X}$, define

$$\theta : \mathsf{VVX} \to \mathsf{uf} \circ \mathsf{l} \circ \mathsf{clpX} : W \mapsto p_W,$$

where $p_W : \mathsf{l}(\mathsf{clpX}) \to 2$ is given on generators by

$$p_W : \mathsf{l} \circ \mathsf{clpX} \to 2 : \Box(\boldsymbol{a}; b) \mapsto \begin{cases} \top & \text{if } W \in \boxdot(\boldsymbol{a}; b) \\ \bot & \text{otherwise} \end{cases}$$

Lemma 4.7. *The assignment θ is well defined.*

Proof. In order to show that θ is well defined, we need to show that p_W is an ultrafilter, that is, a boolean algebra homomorphism $\mathsf{l} \circ \mathsf{clpX} \to 2$. Since l is defined by generators and relations it suffices to show that the images of the generators under p_W satisfy the relations I_1 through I_7. We leave this straightforward verification to the reader. □

The following lemma provides the key ingredient for proving that ξ and θ are continuous and inverses of each other.

Lemma 4.8. *Let $\mathbb{X}$ be a Stone space. We have*

$$W_p \in \boxdot(\boldsymbol{a}; b) \quad \text{if and only if} \quad p(\Box(\boldsymbol{a}; b)) = \top.$$

Proof. If $p(\Box(\boldsymbol{a}; b)) = \bot$, then by construction $W_p \notin \boxdot(\boldsymbol{a}; b)$. So suppose $W_p \notin \boxdot(\boldsymbol{a}, b)$. Then for every witness w of $(\boldsymbol{a}, b)$ there exists (c_w, d_w) which is witnessed by w and is such that $p(\Box(c_w, d_w)) = \bot$.

The collection of witnesses is the set $A = \Diamond a_1 \cap \cdots \cap \Diamond a_n \cap \Box b$. This is a closed set of $V\mathbb{X}$, and it is covered by the collection

$$\{\Diamond c_{w,1} \cap \cdots \cap \Diamond c_{w,m_w} \cap \Box d_w \mid w \in A\}.$$

Clearly this set is an open covering of A, so by compactness of $V\mathbb{X}$ there must be a finite subcover of A. That is

$$A \subseteq \bigcup_{w \in A'} (\Diamond c_{w,1} \cap \cdots \cap \Diamond c_{w,m_w} \cap \Box d_w),$$

where A' is some finite subset of A serving as an index. Now it follows from Lemma 4.9 below that $p(\Box(\boldsymbol{a};b)) = \bot$. $\qquad\square$

The following technical result is motivated by the proof of Lemma 4.8.

Lemma 4.9. *Let* $\mathbb{X}$ *be a Stone space,* $a_i, b, c_j, d \in \mathsf{clp}\mathbb{X}$. *Suppose* $A = \Diamond a_1 \cap \cdots \cap \Diamond a_n \cap \Box b$ *is covered by the finite set* $\{C_i = \Diamond c_{i,1} \cap \cdots \cap \Diamond c_{i,n_i} \cap \Box d_i \mid 1 \leq i \leq m\}$. *Suppose* $p : \mathsf{I} \circ \mathsf{clp}\mathbb{X} \to 2$ *is a ultrafilter and* $p(\Box(c_{i,1}, \ldots, c_{i,n_i}; d_i)) = \bot$ *for all* C_i *in the given cover. Then* $p(\Box(a_1, \ldots, a_n; b)) = \bot$.

Proof. If $A = \emptyset$ the lemma is trivial, so henceforth we shall assume $A \neq \emptyset$.

Part 1. Since (clearly) $b \in A$ there must be a C_i containing b. Call this $C = \Diamond c_1 \cap \cdots \cap \Diamond c_k \cap \Box d$. Now consider $b_j := b \setminus c_j$. This is not in C. If it is not in A, then we must have $c_j \supseteq a_i$ for some i, because clearly $b_j \subseteq b$. If it *is* in A, then it must be in another element of the cover, say, $C_j = \Diamond c_{j1} \cap \cdots \cap \Diamond c_{jn_j} \cap \Box d_j$. Observe that $b_j = b \setminus c_j \subseteq d_j$.

Next, consider $b_{j,k} := b \setminus (c_j \cup c_{jk})$, where $1 \leq k \leq n_j$. If this is not in A, then we must have $a_i \subseteq c_j \cup c_{jk}$ for some i. If it is in A, then it must be in one of the elements of the cover, say, $C_{jk} = \Diamond c_{jk,1} \cap \cdots \cap \Diamond c_{jk,n_{jk}} \cap \Box d_{jk}$. Note that $b_{j,k}$ is not in C and not in C_j by construction. Again, observe that $b_{j,k} \subseteq d_{jk}$. Continuing this way gives a tree, see the diagram below for intuition.

Each $b_{j_1, j_2, \ldots, j_k} = b \setminus (c_{j_1} \cup c_{j_1 j_2} \cup \cdots \cup c_{j_1 j_2 \ldots j_k})$ is in none of the preceding cover elements. Since we started with a finite cover, this process must terminate, i.e. the branches of our tree must be finite. That is, at some point $b_{j_1, j_2, \ldots, j_k}$ is not in A, and since clearly $b_{j_1, j_2, \ldots, j_k} \subseteq b$, it must be the case that $b_{j_1, j_2, \ldots, j_k} \notin \Diamond a_i$ for some a_i, i.e. we must have $a_i \subseteq (c_{j_1} \cup c_{j_1 j_2} \cup \cdots \cup c_{j_1 j_2 \ldots j_k})$.

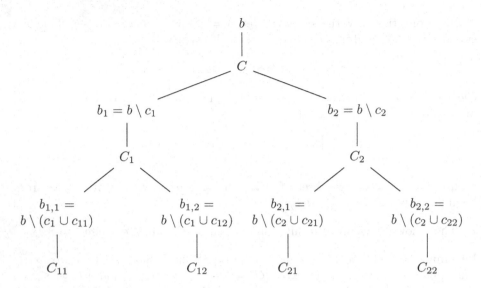

Part 2. Now we have set ourselves up for the proof of the proposition. We will use rule I_5 finitely many times. The first step is:

$$\Box\,(a;b) \leq \Box(a,c_1;b) \vee \Box(a;b \setminus c_1). \tag{3}$$

We continue using I_5 as follows: given an element of the form $\Box(a,c;b_{j_1,j_2,\ldots,j_k})$ we verify what is the lowest entry ℓ such that $c_{j_1 j_2 \cdots j_k \ell}$ is not in c, and apply I_5 using this. It can be seen that (3) above is also obtained in this way. Thus the first two iterations are:

$$\Box(a;b) \leq \Box(a,c_1;b) \qquad\qquad \vee \Box(a;b \setminus c_1)$$
$$\leq \Box(a,c_1,c_2;b) \vee \Box(a,c_1;b_2) \quad \vee \Box(a,c_{11};b_1) \vee \Box(a;b_{1,11})$$

For an entry, we cannot proceed if either all $c_{j_1 \cdots j_k,\ell}$ from a $C_{j_1 \cdots j_k}$ in the tree already occur in c, or if the thing we subtract from b, i.e. $c_{j_1} \cup c_{j_1 j_2} \cup \cdots \cup c_{j_1 j_2 \cdots j_k}$ contains one of the a_i (for then $c_{j_1 j_2 \cdots j_k,\ell}$ is not defined). In the first case, we have

$$\Box(a,c;b \setminus (c_{j_1} \cup c_{j_1 j_2} \cup \cdots \cup c_{j_1 j_2 \cdots j_k}))$$
$$\leq \Box(c_{j_1 \cdots j_k,1}, \ldots, c_{j_1 \cdots j_k, n_{j_1 \cdots j_k}}; d_{j_1 \cdots j_k}). \tag{4}$$

The inequality follows from using I_6 a lot, and applying I_3 to the fact that $b_{j_1,j_2,\ldots,j_k} \subseteq d_{j_1 \cdots j_k}$. As the right hand side of (4) is one of the elements in the cover, we get

$$p(\Box(a,c;b \setminus (c_{j_1} \cup c_{j_1 j_2} \cup \cdots \cup c_{j_1 j_2 \cdots j_k})))$$
$$\leq p(\Box(c_{j_1 \cdots j_k,1}, \ldots, c_{j_1 \cdots j_k, n_{j_1 \cdots j_k}}; d_{j_1 \cdots j_k})) = \bot.$$

In the second case, we get $\bot$ because the intersection of one of the a_i and $b_{j_1 j_2 \cdots j_k}$ is empty, and we use I_4 and I_1.

Since this procedure is finite, this yields $\Box(a;b) \leq \bot$, as desired. $\Box$

As a corollary of Lemma 4.8 we obtain the following lemma.

Lemma 4.10. *The maps θ and ξ are continuous and each others inverses. Hence ξ is a homeomorphism.*

Proof. We first prove continuity. The open subsets of $\mathsf{uf} \circ \mathsf{l} \circ \mathsf{clp}\mathbb{X}$ are generated by $(\!|\Box(\boldsymbol{a};b)|\!) = \{p \mid p(\Box(\boldsymbol{a};b)) = \top\}$, where $(\boldsymbol{a};b) = (a_1, \ldots, a_n; b)$ is an $(n+1)$-tuple of clopen subsets of $\mathbb{X}$. We have

$$\theta^{-1}(\!|\Box(\boldsymbol{a};b)|\!) = \theta^{-1}(\{p \mid p(\Box(\boldsymbol{a};b)) = \top\})$$
$$= \{W \in \mathsf{VV}\mathbb{X} \mid W \in \boxplus(\boldsymbol{a};b)\} = \boxplus(\boldsymbol{a};b),$$

which is clopen in $\mathsf{VV}\mathbb{X}$. Similarly $\xi^{-1}(\boxplus(\boldsymbol{a};b)) = \{p \mid W_p \in \boxplus(\boldsymbol{a};b)\} = (\!|\Box(\boldsymbol{a};b)|\!)$.

We now show prove $\xi \circ \theta$ and $\theta \circ \xi$ are identities. For the former, observe that Lemma 4.8 implies $p(\Box(\boldsymbol{a};b)) = \top$ iff $W_p \in \boxplus(\boldsymbol{a};b)$ iff $p_{W_p}(\Box(\boldsymbol{a};b)) = \top$. So p and p_{W_p} coincide on the generators of $\mathsf{l}(\mathsf{clp}\mathbb{X})$, therefore $p = p_{W_p}$ and hence $\theta \circ \xi = \mathrm{id}_{\mathsf{uf} \circ \mathsf{l} \circ \mathsf{clp}\mathbb{X}}$.

To see that $\xi \circ \theta = \mathrm{id}_{\mathsf{VV}\mathbb{X}}$, note that it suffices to show that W and W_{p_W} are in the same generating opens of the topology. Since the diamond is dual to the box it suffices to show that $W \notin \boxplus(\boldsymbol{a};b)$ iff $W_{p_W} \notin \boxplus(\boldsymbol{a};b)$. This is easy: $W \notin \boxplus(\boldsymbol{a};b)$ iff $p_W(\Box(\boldsymbol{a},b)) = \bot$ iff $W_{p_W} \notin \boxplus(\boldsymbol{a};b)$. The first "iff" holds by definition of θ, the second by Lemma 4.8. $\square$

Each Stone space $\mathbb{X}$ gives rise to a homeomorphism ξ, so we get a transformation $(\xi_\mathbb{X})_{\mathbb{X} \in \mathbf{Stone}} : \mathsf{uf} \circ \mathsf{l} \circ \mathsf{clp} \to \mathsf{VV}$. This transformation is in fact natural:

Lemma 4.11. *The collection $\xi = (\xi_\mathbb{X})_{\mathbb{X} \in \mathbf{Stone}} : \mathsf{uf} \circ \mathsf{l} \circ \mathsf{clp} \to \mathsf{VV}$ is a natural isomorphism.*

Proof. We have already seen that $\xi_\mathbb{X}$ is a homeomorphism for every Stone space $\mathbb{X}$ (i.e. an isomorphism in **Stone**), so it is left to show naturality. That is, where $f : \mathbb{X} \to \mathbb{X}'$ is a continuous function we need to show that

$$
\begin{array}{ccc}
\mathsf{uf} \circ \mathsf{l} \circ \mathsf{clp}\mathbb{X} & \xrightarrow{\;\xi_\mathbb{X}\;} & \mathsf{VV}\mathbb{X} \\
{\scriptstyle \mathsf{uf} \circ \mathsf{l} \circ \mathsf{clp} f}\downarrow & & \downarrow{\scriptstyle \mathsf{VV} f} \\
\mathsf{uf} \circ \mathsf{l} \circ \mathsf{clp}\mathbb{X}' & \xrightarrow{\;\xi_\mathbb{X}\;} & \mathsf{VV}\mathbb{X}'
\end{array}
$$

commutes. Since elements of a Stone space are uniquely determined by the clopen sets in which they are contained, it suffices to show that for all $p \in \mathsf{uf} \circ \mathsf{l} \circ \mathsf{clp}\mathbb{X}$ and $a_i, b \in \mathsf{clp}\mathbb{X}$ we have

$$\mathsf{VV}f(\xi_\mathbb{X}(p)) \in \boxplus(\boldsymbol{a};b) \quad \text{iff} \quad \xi_{\mathbb{X}'}(\mathsf{uf} \circ \mathsf{l} \circ \mathsf{clp}f(p)) \in \boxplus(\boldsymbol{a};b).$$

This follows from a straightforward computation:

$$
\begin{aligned}
\mathsf{VV}f(\xi_X(p)) \in \square(a;b) \quad &\text{iff} \quad \xi_X(p) \in (\mathsf{VV}f)^{-1}(\square(a;b)) \\
&\text{iff} \quad \xi_X(p) \in \square(f^{-1}a; f^{-1}b) \\
&\text{iff} \quad p(\square(f^{-1}a; f^{-1}b)) = \top \\
&\text{iff} \quad p(\mathsf{l} \circ \mathsf{clp}f(\square(a;b))) = \top \\
&\text{iff} \quad \mathsf{uf} \circ \mathsf{l} \circ \mathsf{clp}f(p)(\square(a;b)) = \top \\
&\text{iff} \quad \xi_{X'}(\mathsf{uf} \circ \mathsf{l} \circ \mathsf{clp}f(p)) \in \square(a;b)
\end{aligned}
$$

We conclude that $(\xi_X)_{X \in \mathbf{Stone}}$ is indeed a natural isomorphism. $\square$

As an immediate corollary we obtain the main theorem of this section.

Theorem 4.12. *We have a dual equivalence*

$$\mathbf{Alg}(\mathsf{I}) \equiv^{\mathrm{op}} \mathbf{Coalg}(\mathsf{VV}).$$

As V is dual to the functor M on boolean algebras, we obtain the following corollary which confirms the intuition that INL is "modal logic taken twice".

Corollary 4.13. *The functor I is naturally isomorphic to the composition $\mathsf{M} \circ \mathsf{M}$.*

5   Descriptive Frames as Coalgebras

We show that the descriptive frames from Definition 3.7 are precisely coalgebras for the double Vietoris functor VV.

Theorem 5.1. *We have*

$$\mathbf{D\text{-}INL} \cong \mathbf{Coalg}(\mathsf{VV}).$$

A helpful tool in the proof of Theorem 5.1 is the notion of the *largest representative* of a set $d \subseteq X$ in a general INL-frame (X, N, A).

Definition 5.2. Let (X, N, A) be a general INL-frame and $d \subseteq X$. Then we define the *largest representative* of d to be

$$\overline{d} = \bigcap \{a \in A \mid d \subseteq a\}.$$

This is of course the topological closure of d in the topology on X generated by the clopen base A. It enjoys the following useful properties.

Lemma 5.3. *Let (X, N, A) be a general INL-frame, $d \subseteq X$ and $a_1, \ldots, a_n, b \in A$. Then d witnesses $(a_1, \ldots, a_n; b)$ if and only if $\overline{d}$ does.*

Proof. We show that $d \cap a_i \neq \emptyset$ iff $\overline{d} \cap a_i \neq \emptyset$ and $d \subseteq b$ iff $\overline{d} \subseteq b$. It is easy to see that this proves the lemma. Suppose $d \cap a_i \neq \emptyset$. Since $d \subseteq \overline{d}$ we also have $\overline{d} \cap a_i \neq \emptyset$. Conversely, if $d \cap a_i = \emptyset$ then $d \subseteq X \setminus a_i$ and since the latter is in A we have $\overline{d} \subseteq X \setminus a_i$. This implies $\overline{d} \cap a_i = \emptyset$. Next suppose $d \subseteq b$, then by definition $\overline{d} \subseteq b$, because $b \in A$. Conversely, if $\overline{d} \subseteq b$ we have $d \subseteq \overline{d} \subseteq b$. $\square$

Lemma 5.4. *Let (X, N, A) be a descriptive INL-frame, $d \subseteq X$ and $x \in X$. Then $d \in N(x)$ if and only if $\overline{d} \in N(x)$.*

Proof. This follows directly from the proof of Lemma 5.3.   □

The following two lemmas describe the object part of the isomorphism from Theorem 5.1.

Lemma 5.5. *Let $(\mathbb{X}, \gamma)$ be a VV-coalgebra. Write X for the space underlying $\mathbb{X}$ and let $N_\gamma(x) = \{d \subseteq X \mid \overline{d} \in \gamma(x)\}$. Then $(X, N_\gamma, \mathsf{clp}\mathbb{X})$ is a descriptive INL frame.*

Proof. We know that $\mathsf{clp}X$ is closed under boolean operations and it follows from continuity of γ that $\mathsf{clp}X$ is closed under $m_\square$. Furthermore, $(X, N_\gamma, \mathsf{clp}\mathbb{X})$ is differentiated because $\mathbb{X}$ is Hausdorff and compact because $\mathbb{X}$ is compact.

Lastly, we show that it is crowded. Suppose $c \notin N_\gamma(x)$. Without loss of generality we may assume c to be closed, hence an element of $\mathsf{V}\mathbb{X}$, because we know from Lemma 5.3 that c and $\overline{c}$ witness precisely the same tuples. It follows from the definition of N_γ that $c \notin \gamma(x)$. Since $\gamma(x)$ is a closed subset of $\mathsf{V}\mathbb{X}$, there must be a basic clopen $\Diamond a_1 \cap \cdots \cap \Diamond a_n \cap \square b$ containing c and disjoint from $\gamma(x)$. Therefore c witnesses $(a_1, \ldots, a_n; b)$ while none of the elements in $\gamma(x)$ witness $(a_1, \ldots, a_n; b)$. It then follows from the definition of N_γ and Lemma 5.3 that none of the $d \in N_\gamma(x)$ witness $(a_1, \ldots, a_n; b)$. Therefore $(X, N_\gamma, \mathsf{clp}\mathbb{X})$ is crowded.   □

Lemma 5.6. *Let (X, N, A) be a descriptive INL-frame, write $\mathbb{X}$ for the set X topologized by the clopen subbase A and let $\gamma_N : \mathbb{X} \to \mathsf{VV}\mathbb{X} : x \mapsto \{c \in \mathsf{V}\mathbb{X} \mid c \in N(x)\}$. Then $(\mathbb{X}, \gamma_N)$ is a VV-coalgebra.*

Proof. The topological space $\mathbb{X}$ is zero-dimensional because A is closed under complementation (hence is a *clopen* base). Moreover, $\mathbb{X}$ is compact Hausdorff because (X, N, A) is compact and differentiated, so $\mathbb{X}$ is a Stone space.

In order to show that γ_N is well defined, we need to show that $\gamma_N(x)$ is a closed subset of $\mathsf{V}\mathbb{X}$ for every $x \in \mathbb{X}$. Suppose $c \in \mathsf{V}\mathbb{X}$ and $c \notin \gamma_N(x)$. Then $c \notin N(x)$, and because (X, N, A) is crowded we can find $(a_1, \ldots, a_n; b)$ which is witnessed by c but by none of the elements in $N(x)$. This implies $c \in \Diamond a_1 \cap \cdots \cap \Diamond a_n \cap \square b$ and $\Diamond a_1 \cap \cdots \cap \Diamond a_n \cap \square b$ is disjoint from $\gamma_N(x)$. Thus we have found an open neighbourhood of c disjoint from $\gamma_N(x)$ so $\gamma_N(x)$ is closed in $\mathsf{V}\mathbb{X}$.

For continuity of γ_N, it suffices to show that $\gamma_N^{-1}(\square(a_1, \ldots, a_n; b))$ is clopen in $\mathbb{X}$ for all $a_1, \ldots, a_n, b \in A$. This is a consequence of the fact that A is closed under $m_\square$, because

$$\gamma_N^{-1}(\square(a_1, \ldots, a_n; b)) = m_\square(a_1, \ldots, a_n; b).$$

We conclude that $(\mathbb{X}, \gamma_N)$ is a VV-coalgebra.   □

We proceed with the proof of Theorem 5.1.

Proof of Theorem 5.1. First we verify that the assignments from Lemmas 5.5 and 5.6 define a bijection between descriptive INL-frames and VV-coalgebras. Let $(\mathbb{X}, \gamma)$ be a VV-coalgebra. Lemma 5.5 assigns to this the descriptive INL-frame $(X, N_\gamma, \mathsf{clp}\mathbb{X})$. We know that the topology on X generated by $\mathsf{clp}\mathbb{X}$ yields the topological space $\mathbb{X}$, so applying Lemma 5.6 to $(X, N_\gamma, \mathsf{clp}\mathbb{X})$ yields the VV-coalgebra $(\mathbb{X}, \gamma_{N_\gamma})$. Furthermore, for a closed set $c \in V\mathbb{X}$ we have $c \in \gamma_{N_\gamma}(x)$ iff $c \in N_\gamma(x)$ iff $c \in \gamma(x)$, hence $\gamma = \gamma_{N_\gamma}$ and $(\mathbb{X}, \gamma) = (\mathbb{X}, \gamma_{N_\gamma})$.

Conversely, suppose given a descriptive INL-frame (X, N, A). Write τ_A for the topology on X generated by the (clopen) base A and let $\mathbb{X} = (X, \tau_A)$. Then Lemma 5.6 sends (X, N, A) to $(\mathbb{X}, \gamma_N)$, which is in turn send to $(X, N_{\gamma_N}, \mathsf{clp}\mathbb{X})$ by Lemma 5.5. We know that the clopen sets of τ_A are precisely the sets in A, so $\mathsf{clp}\mathbb{X} = A$. Comparing the neighbourhood functions gives

$$
\begin{aligned}
d \in N(x) \quad &\text{iff} \quad \overline{d} \in N(x) &&\text{(Lemma 5.4)}\\
&\text{iff} \quad \overline{d} \in \gamma_N(x) &&\text{(Lemma 5.6)}\\
&\text{iff} \quad \overline{d} \in N_{\gamma_N}(x) &&\text{(Lemma 5.5)}\\
&\text{iff} \quad d \in N_{\gamma_N}(x) &&\text{(Lemma 5.4)}
\end{aligned}
$$

and therefore $(X, N, A) = (X, N_{\gamma_N}, \mathsf{clp}\mathbb{X})$. This proves the isomorphism on objects.

Let $(\mathbb{X}, \gamma)$ and $(\mathbb{X}', \gamma')$ be two VV-coalgebras and $f : X \to X'$ a function. We claim that f is a VV-coalgebra morphism if and only if it is a general INL-morphism. If f is a general INL-morphism then clearly it is continuous. Since it is an INL-morphism moreover the diagram

$$
\begin{array}{ccc}
X & \xrightarrow{\;f\;} & X' \\
\downarrow{\scriptstyle N_\gamma} & & \downarrow{\scriptstyle N_{\gamma'}} \\
\mathsf{PP}X & \xrightarrow{\;\mathsf{PP}f\;} & \mathsf{PP}X'
\end{array}
\tag{5}
$$

commutes. It follows immediately that

$$
\begin{array}{ccc}
X & \xrightarrow{\;f\;} & X' \\
\downarrow{\scriptstyle \gamma} & & \downarrow{\scriptstyle \gamma'} \\
\mathsf{VV}X & \xrightarrow{\;\mathsf{VV}f\;} & \mathsf{VV}X'
\end{array}
\tag{6}
$$

commutes, so f is an VV-coalgebra morphism.

Conversely, if f is continuous and (6) commutes, then (5) commutes because $d \in N_{\gamma'}(f(x))$ iff $\overline{d} \in N_{\gamma'}(f(x))$ iff $\overline{d} \in \gamma'(f(x))$ iff $\overline{d} \in \mathsf{VV}f(\gamma(x))$ iff $\overline{d} \in \mathsf{PP}f(N_\gamma(x))$ iff $d \in \mathsf{PP}f(N_\gamma(x))$. The last "iff" follows from Lemma 5.4. It is a general INL-morphism because continuity implies that f^{-1} sends admissible subsets to admissible subsets. $\square$

We have now completed the strategy outlined in diagram (1). As a corollary we obtain Theorem 3.9, whose formulation we copy here for the reader's convenience.

Theorem 3.9. *We have a dual equivalence*

$$\textbf{D-INL} \equiv^{\text{op}} \textbf{BAIO}.$$

Remark 5.7. Careful inspection of the definitions shows that the duality in Theorem 3.9 is given on objects by the construction in Example 3.6 and the subsequent paragraph.

6 Applications

We will give two applications of our results for completeness of INL-based logic and theory of INL-bisimulations.

6.1 Completeness

An *extension* of INL is any set of INL-formulas which contains INL (that is, all the INL-formulas valid on all INL-frames) and is closed under the rules of Modus Ponens and (RE). The latter states that for formulas α, β and φ, if $\alpha \leftrightarrow \beta$, then $\varphi[\alpha/\beta]$ holds, where $\varphi[\alpha/\beta]$ is the result of possibly replacing some occurrences of α in φ by β (see [3]).

It is well known that every modal logic is sound and complete with respect to its algebraic semantics [7,10,22]. From the main completeness result of [3] it follows that BAIOs provide an algebraic semantics for INL. Then the standard argument yields that every extension L of INL is sound and complete with respect to the class of BAIOs validating L. Moreover, as a direct corollary of Theorem 3.9, we obtain:

Theorem 6.1. *Every extension of INL is sound and complete with respect to descriptive INL-frames.*

6.2 Bisimulations

We briefly discuss bisimulations for INL and derive a Hennessy-Milner property for descriptive INL-frames. We work in a setting without proposition letters, but all results carry over to the setting *with* proposition letters.

Recall the definition of a bisimulation, in its coalgebraic form:

Definition 6.2. Let (X, N) and (X', N') be neighbourhood frames. A relation $B \subseteq X \times X'$ is an *INL-bisimulation* if there exists a neighbourhood function $M : B \to \text{PP}B$ such that

$$
\begin{array}{ccccc}
X & \xleftarrow{\ \pi\ } & B & \xrightarrow{\ \pi'\ } & X' \\
\downarrow{\scriptstyle N} & & \downarrow{\scriptstyle M} & & \downarrow{\scriptstyle N'} \\
\text{PP}X & \xleftarrow[\text{PP}\pi]{} & \text{PP}B & \xrightarrow[\text{PP}\pi']{} & \text{PP}X'
\end{array}
$$

commutes. Two states are called bisimilar if they are linked by a bisimulation.

Since PP weakly preserves pullbacks, bisimilarity and behavioural equivalence coincide. In particular, if $(X, N) \xrightarrow{f} (Z, M) \xleftarrow{f'} (X', N')$ is a cospan in **INL** (witnessing behavioural equivalence of some states), then the pullback of f and f' in **Set** is a bisimulation.

Define bisimulations between descriptive frames as follows:

Definition 6.3. A *descriptive INL-bisimulation* between $(\mathbb{X}, \gamma)$ and $(\mathbb{X}', \gamma')$ is a subspace $B \subseteq \mathbb{X} \times \mathbb{X}'$ such that B is a bisimulation between the underlying neighbourhood frames.

In [11] the notion of Λ-bisimulation is introduced, where Λ is a so-called (characteristic) modal signature for an endofunctor on **Stone**. It is straightforward to see that the interpretation of INL in descriptive INL-frames can be translated to the setting used in *op. cit.* Moreover, an easy computation shows that every descriptive INL-bisimulation in the sense of Definition 6.3 is a Λ-bisimulation. We expect that the converse holds as well but at present do not have a proof of this. We leave it as an interesting open question.

We now prove a Hennessey-Milner property for descriptive INL-frames. This is the INL analogue of [5, Corollary 3.9].

Theorem 6.4. *Let $(\mathbb{X}, \gamma)$ and $(\mathbb{X}', \gamma')$ be descriptive frames. Then $x \in \mathbb{X}$ and $x' \in \mathbb{X}'$ are logically equivalent iff they are behaviourally equivalent iff they are linked by a descriptive INL-bisimulation.*

Proof. Logical equivalence implies behavioural equivalence because every two logically equivalent states are identified by the theory map to the canonical model (i.e. the final object in **D-INL**). If x and x' are behaviourally equivalent then there are morphisms in **D-INL** such that $f(x) = f'(x')$. The pullback of f and f' viewed as functions in **Set** is a bisimulation between the underlying neighbourhood frames (by the text following Definition 6.2). Moreover, this pullback is closed in $\mathbb{X} \times \mathbb{X}'$ because pullbacks in **Stone** are computed as in **Set**. Hence behavioural equivalence implies bisimilarity. Lastly, bisimilarity implies logical equivalence by design. $\square$

We can apply this theorem to all INL-frames that carry a descriptive structure, that is, to $(X, N) \in$ **INL** such that there exists A making (X, N, A) a descriptive INL-frame.

Corollary 6.5. *Suppose the INL-frames (X, N) and (X', N') both carry a descriptive structure. Then between these frames, bisimilarity coincides with logical equivalence.*

Restricting this corollary entails [3, Theorem 3.1], the Hennessy-Milner property for finite frames, because all finite neighbourhood frames carry a descriptive frame structure (see Proposition 3.8).

Theorem 6.6. *Let (X, N) and (X', N') be finite neighbourhood frames. Then bisimilarity coincides with logical equivalence.*

7 Conclusion and Future Work

In this paper we introduced descriptive frames for Instantial Neighbourhood Logic (INL) and showed that these frames are dual to BAIOs, the algebras for INL. Coalgebra provided a key for obtaining this duality. We first presented BAIOs as algebras for the functor I on the category of boolean algebras. We also represented descriptive INL-frames as coalgebras for the double Vietoris functor VV on the category of Stone spaces. Finally, we showed that the category of I-algebras is dual to the category of VV-coalgebras, leading to the desired duality result. As a corollary we obtained that every extension of INL is sound and complete with respect to descriptive INL-frames. One interesting question for future work is whether one can obtain an analogue of the celebrated Sahlqvist completeness and correspondence result for extensions of INL.

We recall that the Vietoris functor V is dual to the functor M on the category of boolean algebras (Sect. 2). Intuitively M freely adds one layer of normal modalities to a boolean algebra. We showed in this paper that BAIOs can be represented as algebras for M∘M (Sect. 4). Therefore, BAIOs can be seen as "modal algebras squared" and INL itself is, in a way, "the basic modal logic squared". This provokes the question of what "modal algebras cubed" looks like, i.e. what logic and algebras correspond to the functor M ∘ M ∘ M, and similar questions for the n-fold composition of M.

Recall that monotone neighbourhood logic EM algebraically corresponds to a functor N on boolean algebras which intuitively adds one layer of monotone modalities to a boolean algebra [17,18]. This generates a question whether NN-algebras and the corresponding logic admit an "INL-style axiomatization".

Another related formalism that would be interesting to investigate is that of *positive* INL. The algebras for this logic are *distributive lattices with instantial operators* (DLIOs):

Definition 7.1. A *distributive lattice with instantial operators* (DLIO) is a tuple $(D, (f_n)_{n \in \omega}, (g_n)_{n \in \omega})$ consisting of a distributive lattice D and two collections of ω-indexed maps $f_n, g_n : D^{n+1} \to D$ such that: (1) The f_n satisfy (B_1) to (B_7) from Definition 3.1, where, in absence of negation, we reformulate (B_5) as

$$f_n(a_1, \ldots, a_n; b) \leq f_{n+1}(a_1, \ldots, a_n, c; b) \vee f_n(a_1, \ldots, a_n; b \wedge d),$$

whenever $c \vee d = \top$;   (2) The g_n satisfy relations dual to the ones for f_n;   and (3) The f_n and g_n satisfy the duality axioms

$$g_{n+1}(a_1, \ldots, a_n, b'; b) \wedge \bigwedge_{i=1}^{m} g_n(a_1, \ldots, a_n; a_i' \vee b)$$
$$\leq g_n(a_1, \ldots, a_n; b) \vee f_m(a_1', \ldots, a_m'; b') \tag{D_1}$$

and

$$f_n(a_1, \ldots, a_n; b) \wedge g_m(a_1', \ldots, a_m'; b')$$
$$\leq f_{n+1}(a_1, \ldots, a_n, b'; b) \vee \bigvee_{i=1}^{m} f_n(a_1, \ldots, a_n; a_i' \wedge b). \tag{D_2}$$

These are of course algebras for a endofunctor J on the category **DL** of distributive lattices and homomorphisms. In analogy with the results of this paper, one would expect that descriptive frames for positive INL are isomorphic to coalgebras for the double convex Vietoris functor V_c on the category of Priestley spaces, as this is the Priestley space analogue of the Vietoris functor [6,27,33]. We expect the following duality result:

Conjecture 7.2. *We have a dual equivalence*

$$\mathbf{Alg}(J) \equiv^{op} \mathbf{Coalg}(V_cV_c).$$

Finally, related to positive INL, an interesting question is to consider the geometric logic analogue of INL and to verify a slogan of [4], which in the case of INL will read as

$$\text{Geometric INL} = \text{Positive INL} + \text{Scott continuity}.$$

If correct, this may also provide a novel algebraic presentation of the double Vietoris powerlocale studied extensively by Vickers [34–36].

Acknowledgements. The authors would like to thank the reviewers of CMCS 2020 for their helpful comments, which improved the presentation of the paper.

References

1. van Benthem, J., Bezhanishvili, N., Enqvist, S.: A new game equivalence, its logic and algebra. J. Philos. Logic **48**(4), 649–684 (2019)
2. van Benthem, J., Bezhanishvili, N., Enqvist, S.: A propositional dynamic logic for instantial neighbourhood semantics. Studia Logica **107**(4), 719–751 (2019)
3. van Benthem, J., Bezhanishvili, N., Enqvist, S., Yu, J.: Instantial neighbourhood logic. Rev. Symbol. Logic **10**(1), 116–144 (2017)
4. Bezhanishvili, N., de Groot, J., Venema, Y.: Coalgebraic geometric logic. In: Roggenbach, M., Sokolova, A. (eds.) Proceedings of CALCO 2019, Dagstuhl, Germany, pp. 7:1–7:18 (2019). Schloss Dagstuhl-Leibniz-Zentrum fuer Informatik
5. Bezhanishvili, N., Fontaine, G., Venema, Y.: Vietoris bisimulations. J. Logic Comput. **20**(5), 1017–1040 (2010)
6. Bezhanishvili, N., Kurz, A.: Free modal algebras: a coalgebraic perspective. In: Mossakowski, T., Montanari, U., Haveraaen, M. (eds.) CALCO 2007. LNCS, vol. 4624, pp. 143–157. Springer, Heidelberg (2007). https://doi.org/10.1007/978-3-540-73859-6_10
7. Blackburn, P., de Rijke, M., Venema, Y.: Modal Logic. Cambridge Tracts in Theoretical Computer Science. Cambridge University Press, Cambridge (2001)
8. Celani, S., Jansana, R.: Priestley duality, a Sahlqvist theorem and a Goldblatt-Thomason theorem for positive modal logic. Logic J. IGPL **7**, 683–715 (1999). 12
9. Idrees, S.K., Fanfakh, A.B.M.: Performance and energy consumption prediction of randomly selected nodes in heterogeneous cluster. In: Al-mamory, S.O., Alwan, J.K., Hussein, A.D. (eds.) NTICT 2018. CCIS, vol. 938, pp. 21–34. Springer, Cham (2018). https://doi.org/10.1007/978-3-030-01653-1_2

10. Chagrov, A., Zakharyaschev, M.: Modal Logic. Oxford University Press, Oxford (1997)
11. Enqvist, S., Sourabh, S.: Bisimulations for coalgebras on Stone spaces. J. Logic Comput. **28**(6), 991–1010 (2018)
12. Esakia, L.L.: Topological Kripke models. Soviet Mathematics Doklady **15**, 147–151 (1974)
13. Esakia, L.L.: The problem of dualism in the intuitionistic logic and Browerian lattices. In: V International Congress of Logic, Methodology and Philosophy of Science, Canada, pp. 7–8 (1975)
14. Esakia, L.: Heyting Algebras. Trends in Logic, Bezhanishvili, G., Holliday, W.H. (eds.) vol. 50. Springer, Cham (2019). https://doi.org/10.1007/978-3-030-12096-2. Translated by A. Evseev
15. Furber, R., Kozen, D., Larsen, K.G., Mardare, R., Panangaden, P.: Unrestricted stone duality for Markov processes. In: Proceedings of LICS 2017, Reykjavík, Iceland, pp. 1–9. IEEE Press (2017)
16. Goldblatt, R.I.: Metamathematics of modal logic I. Rep. Math. Logic **6**, 41–78 (1976)
17. Hansen, H.H.: Monotonic modal logics. Master's thesis, Institute for Logic, Language and Computation, University of Amsterdam (2003)
18. Hansen, H.H., Kupke, C.: A coalgebraic perspective on monotone modal logic. Electron. Notes Theor. Comput. Sci. **106**, 121–143 (2004)
19. Johnstone, P.T.: Stone Spaces. Cambridge Studies in Advanced Mathematics. Cambridge University Press, Cambridge (1982)
20. Jónsson, B., Tarski, A.: Boolean algebras with operators, Part I. Am. J. Math. **73**, 891–939 (1951)
21. Kapulkin, K., Kurz, A., Velebil, J.: Expressiveness of positive coalgebraic logic. In: Bolander, T., Braüner, T., Ghilardi, S., Moss, L.S. (eds.) Proceedings of AIML 2012, pp. 368–385. College Publications (2012)
22. Kupke, C., Kurz, A., Pattinson, D.: Algebraic semantics for coalgebraic logics. In: Proceedings of the Workshop on Coalgebraic Methods in Computer Science (CMCS). Electronic Notes in Theoretical Computer Science, vol. 106, pp. 219–241 (2004)
23. Kupke, C., Kurz, A., Venema, Y.: Stone coalgebras. Theoret. Comput. Sci. **327**(1), 109–134 (2004). Selected Papers of CMCS '03
24. Kupke, C., Pattinson, D.: Coalgebraic semantics of modal logics: an overview. Theoret. Comput. Sci. **412**(38), 5070–5094 (2011). CMCS Tenth Anniversary Meeting
25. Mio, M., Furber, R., Mardare, R.: Riesz modal logic for Markov processes. In: Proceedings of LICS 2017, pp. 1–12, 06 2017
26. Moss, L.S.: Coalgebraic logic. Ann. Pure Appl. Logic **96**(1), 277–317 (1999)
27. Palmigiano, A.: A coalgebraic view on positive modal logic. Theoret. Comput. Sci. **327**, 175–195 (2004)
28. Priestley, H.A.: Representation of distributive lattices by means of ordered Stone spaces. Bull. Lond. Math. Soc. **2**(2), 186–190 (1970)
29. Rutten, J.J.M.M.: Universal coalgebra: a theory of systems. Theoret. Comput. Sci. **249**(1), 3–80 (2000)
30. Sambin, G., Vaccaro, V.: A new proof of Sahlqvist's theorem on modal definability and completeness. J. Symbolic Logic **54**(3), 992–999 (1989)
31. Stone, M.H.: The theory of representations for Boolean algebras. Trans. Am. Math. Soc. **40**, 37–111 (1936)
32. Tuyt, O.: Canonical rules on neighbourhood frames. Master's thesis, ILLC, University of Amsterdam (2016)

33. Venema, Y., Vosmaer, J.: Modal logic and the Vietoris lunctor. In: Bezhanishvili, G. (ed.) Leo Esakia on Duality in Modal and Intuitionistic Logics. OCL, vol. 4, pp. 119–153. Springer, Dordrecht (2014). https://doi.org/10.1007/978-94-017-8860-1_6
34. Vickers, S.: Constructive points of powerlocales. Math. Proc. Cambridge Philos. Soc. **122**(2), 207–222 (1997)
35. Vickers, S.J.: The double powerlocale and exponentiation: a case study in geometric logic. Theory Appl. Categories **12**(13), 272–422 (2004)
36. Vickers, S.J., Townsend, C.F.: A universal characterization of the double power-locale. Theoret. Comput. Sci. **316**(1), 297–321 (2004). Recent Developments in Domain Theory: A collection of papers in honour of Dana S. Scott
37. Vietoris, L.: Bereiche zweiter Ordnung. Monatshefte für Mathematik und Physik **32**(1), 258–280 (1922). https://doi.org/10.1007/BF01696886
38. Yosida, K.: On the representation of vector lattices. Proc. Imperial Acad. **18**(7), 339–342 (1942)
39. Yu, J.: A tableau system for instantial neighborhood logic. In: Artemov, S., Nerode, A. (eds.) LFCS 2018. LNCS, vol. 10703, pp. 337–353. Springer, Cham (2018). https://doi.org/10.1007/978-3-319-72056-2_21
40. Yu, J.: Lyndon interpolation theorem of instantial neighborhood logic - constructively via a sequent calculus. Ann. Pure Appl. Logic **171**(1), 102721 (2020)

Free-Algebra Functors from a Coalgebraic Perspective

H. Peter Gumm[✉]

Philipps-Universität Marburg, Marburg, Germany
gumm@mathematik.uni-marburg.de

Abstract. We continue our study of free-algebra functors from a coal-gebraic perspective as begun in [8]. Given a set Σ of equations and a set X of variables, let $F_\Sigma(X)$ be the free Σ−algebra over X and $\mathcal{V}(\Sigma)$ the variety of all algebras satisfying Σ. We consider the question, under which conditions the *Set*-functor F_Σ weakly preserves pullbacks, kernel pairs, or preimages [9].

We first generalize a joint result with our former student Ch. Henkel, asserting that an arbitrary *Set*−endofunctor F weakly preserves kernel pairs if and only if it weakly preserves pullbacks of epis.

By slightly extending the notion of derivative Σ' of a set of equations Σ as defined by Dent, Kearnes and Szendrei in [3], we show that a functor F_Σ (weakly) preserves preimages if and only if Σ implies its own derivative, i.e. $\Sigma \vdash \Sigma'$, which amounts to saying that weak independence implies independence for each variable occurrence in a term of $\mathcal{V}(\Sigma)$. As a corollary, we obtain that the free-algebra functor will never preserve preimages when $\mathcal{V}(\Sigma)$ is congruence modular.

Regarding preservation of kernel pairs, we show that for n-permutable varieties $\mathcal{V}(\Sigma)$, the functor F_Σ weakly preserves kernel pairs if and only if $\mathcal{V}(\Sigma)$ is a Mal'cev variety, i.e. 2-permutable.

1 Introduction

In his groundbreaking monograph "Universal Coalgebra – a theory of systems" [15] Jan Rutten demonstrated how all sorts of state based systems could be unified under the roof of one abstract concept, that of a coalgebra. Concrete system types – automata, transition systems, nondeterministic, weighted, probabilistic or second order systems – can be modeled by choosing an appropriate functor $F : Set \to Set$ which provides a *type* for the concrete coalgebras just as, on a less abstract level, signatures describe types of algebras.

The (co)algebraic properties of the category Set_F of all F−coalgebras are very much dependent on certain preservation properties of the functor F. A particular property of F, which has been considered relevant from the beginning was the preservation of certain weak limits. In Rutten's original treatise [16], lemmas and theorems were marked with an asterisk, if they used the additional assumption that F *weakly preserves pullbacks*, that is F transforms pullback diagrams in *Set* into weak pullback diagrams.

© IFIP International Federation for Information Processing 2020
Published by Springer Nature Switzerland AG 2020
D. Petrişan and J. Rot (Eds.): CMCS 2020, LNCS 12094, pp. 55–67, 2020.
https://doi.org/10.1007/978-3-030-57201-3_4

In our lecture notes [6], we were able to remove a large number of asterisks from Rutten's original presentation, and in joint work with T. Schröder [9], we managed to split the mentioned preservation condition into two separate conditions, *weak preservation of kernel pairs* and *preservation of preimages*. These properties were then studied separately with an eye on their structure theoretic significance.

It is therefore relevant and interesting to classify *Set* functors according to the mentioned preservation properties.

We start this paper by showing that for arbitrary *Set* functors weak preservation of kernel pairs is equivalent to weak preservation of pullbacks of epis, a result, which was obtained jointly with our former master student Ch. Henkel.

Subsequently, we investigate *Set* functors F_Σ which associate to a set X the free $\Sigma-$algebra over X. It turns out that (weak) preservation of preimages by F_Σ can be characterized utilizing the *derivative* Σ' of Σ, which has been studied a few years ago by Dent, Kearnes, and Szendrei [3]. For arbitrary sets of *idempotent* equations Σ, they show that the variety $\mathcal{V}(\Sigma)$ is congruence modular if and only if $\Sigma \cup \Sigma'$ is inconsistent. Below, we extend their notion of derivative to arbitrary sets of equations (not necessarily idempotent) and are able to show that F_Σ weakly preserves preimages if and only if $\Sigma \vdash \Sigma'$.

Regarding preservation of kernel pairs, we exhibit an algebraic condition, which to our knowledge has not been studied before and which appears to be interesting in its own right. If F_Σ weakly preserves kernel pairs, that is if F_Σ weakly preserves pullbacks of epis, then for any pair p, q of ternary terms satisfying

$$p(x, x, y) \approx q(x, y, y)$$

there exists a quaternary term s such that

$$p(x, y, z) \approx s(x, y, z, z)$$
$$q(x, y, z) \approx s(x, x, y, z).$$

Applying this to the description of $n-$permutable varieties given by Hagemann and Mitschke [10], we find that for an $n-$permutable variety $\mathcal{V}(\Sigma)$, the functor F_Σ weakly preserves kernel pairs if and only if $\mathcal{V}(\Sigma)$ is a Mal'cev variety, i.e. there exists a term $m(x, y, z)$ such that the equations

$$m(x, y, y) \approx x$$
$$m(x, x, y) \approx y$$

are satisfied.

2   Preliminaries

For the remainder of this work we shall denote function application by juxtaposition, associating to the right, i.e. fx denotes $f(x)$ and $f\,g\,x$ denotes $f(g(x))$.

If F is a functor, we denote by $F(X)$ the application of F to an object X and by Ff the application of F to a morphism f.

Given a Set-functor F, an F-coalgebra is simply a map $\alpha : A \to F(A)$, where A is called the *base set* and α the *structure map* of the coalgebra $\mathcal{A} = (A, \alpha)$. A *homomorphism* between coalgebras $\mathcal{A} = (A, \alpha)$ and $\mathcal{B} = (B, \beta)$ is a map $\varphi : A \to B$ satisfying

$$\beta \circ \varphi = F\varphi \circ \alpha.$$

The functor F is called the *type* of the coalgebra. The class of all coalgebras of type F together with their homomorphisms forms a category Set_F. The structure of this category is known to depend heavily on several pullback preservation properties of the functor F.

A pullback diagram

$$
\begin{array}{ccc}
A_1 & \xrightarrow{\ f_1\ } & C \\
{\scriptstyle p_1}\big\uparrow & & \big\uparrow{\scriptstyle f_2} \\
P & \xrightarrow[\ p_2\ ]{} & A_2
\end{array}
$$

is called a *kernel pair*, if $f_1 = f_2$ and it is called a *preimage* if f_1 or f_2 is mono.

A functor F is said to *weakly preserve pullbacks* if F transforms each pullback diagram into a weak pullback diagram. F weakly preserving kernel pairs, resp. preimages are defined likewise.

In order to check whether in the category Set a diagram as above is a weak pullback, we may argue elementwise: (P, p_1, p_2) is a weak pullback, if for each pair (a_1, a_2) with $a_i \in A_i$ and $f_1 a_1 = f_2 a_2$ there exists some $a \in P$ such that $p_1 a = a_1$ and $p_2 a = a_2$.

Hence, to see that a Set-functor F weakly preserves a pullback (P, p_1, p_2), we must check that for any pair (u_1, u_2) with $u_i \in F(A_i)$ and $(Ff_1)u_1 = (Ff_2)u_2$ we can find some $w \in F(P)$ such that $(Fp_1)w = u_1$ and $(Fp_2)w = u_2$.

One easily checks that a *weak preimage* is automatically a *preimage*. Hence, if F preserves monos, then F *preserves* preimages if and only if F *weakly preserves* preimages.

Assuming the axiom of choice, all epis in the category Set are right invertible, hence F preserves epis. Monos are left-invertible, except for the empty mappings $\emptyset_X : \emptyset \to X$ when $X \neq \emptyset$. Hence F surely preserves monos with nonempty domain.

Most Set-functors also preserve monos with empty domain. This will, in particular, be the case for the free-algebra functor F_Σ, which we shall study in the later parts of this work.

For Set-functors F which fail to preserve monos with empty domain, there is an easy fix, modifying F solely on the empty set $\emptyset$ and on the empty mappings $\emptyset_X : \emptyset \to X$, so that the resulting functor $F^\star$ preserves all monos. The details can be found in [1] or in [7]. This modification is irrelevant as far as coalgebras are concerned, since it affects only the empty coalgebra. Yet it allows us to assume from now on, that F preserves all monos and all epis.

If f_1 and f_2 are both injective, then their pullback is called an intersection. It is well known from [18] that a set functor automatically preserves nonempty intersections, and, after possibly modifying it at $\emptyset$ as indicated above, preserves all finite intersections.

3 Weak Preservation of Epi Pullbacks

The following lemma from [9] shows that weak pullback preservation can be split into two separate preservation requirements.

Lemma 1. *For a Set-functor F the following are equivalent:*

1. *F weakly preserves pullbacks.*
2. *F weakly preserves kernels and preimages.*

A special case of a preimage is obtained if we consider a subset $U \subseteq A$ as the preimage of $\{1\}$ along its characteristic function $\chi_U : A \to \{0,1\}$.

$$
\begin{array}{ccc}
A & \xrightarrow{\ \chi_U\ } & \{0,1\} \\
\uparrow & & \uparrow \\
U & \xrightarrow{\ !_U\ } & \{1\}
\end{array}
$$

Such preimages are called *classifying*, and we shall later make use of the following lemma from [9]:

Lemma 2. *A Set-functor F preserves preimages if and only if it preserves classifying preimages.*

In the following section we need to consider the action of a functor on pullback diagrams where both f_1 and f_2 are surjective. Before stating this, we shall prove a useful lemma, which is true in every category:

Lemma 3. *Let morphisms $f : A \to C$ and $f_i : A_i \to C$, for $i = 1,2$ be given, as well as $e_i : A_i \to A$ with left inverses $h_i : A \to A_i$ such that the diagram below commutes. If (K, π_1, π_2) is a weak kernel of f then $(K, h_1 \circ \pi_1, h_2 \circ \pi_2)$ is a weak pullback of f_1 and f_2.*

$$
\begin{array}{ccccc}
 & & A_i & & \\
 & {\scriptstyle h_i}\uparrow\downarrow{\scriptstyle e_i} & & \searrow^{f_i} & \\
K & \underset{\pi_2}{\overset{\pi_1}{\rightrightarrows}} & A & \xrightarrow{\ f\ } & C
\end{array}
$$

Proof. Assuming (K, π_1, π_2) is a weak kernel of f, then setting $k_i := h_i \circ \pi_i$ we obtain

$$f_1 \circ k_1 = f_1 \circ h_1 \circ \pi_1 = f \circ \pi_1 = f \circ \pi_2 = f_2 \circ h_2 \circ \pi_2 = f_2 \circ k_2.$$

This shows that (K, k_1, k_2) is a candidate for a pullback of f_1 with f_2. Let (Q, q_1, q_2) be another candidate, i.e.

$$f_1 \circ q_1 = f_2 \circ q_2,$$

then we obtain

$$f \circ e_1 \circ q_1 = f_1 \circ q_1 = f_2 \circ q_2 = f \circ e_2 \circ q_2,$$

which demonstrates that $(Q, e_1 \circ q_1, e_2 \circ q_2)$ is a competitor to (K, π_1, π_2) for being a weak kernel of f. This yields a morphism $q : Q \to K$ with

$$\pi_i \circ q = e_i \circ q_i.$$

From this we obtain

$$k_i \circ q = h_i \circ \pi_i \circ q = h_i \circ e_i \circ q_i = q_i$$

as required.

Theorem 4. *Let C be a category with finite sums and kernel pairs. If a functor $F : C \to C$ weakly preserves kernel pairs, then it weakly preserves pullbacks of retractions.*

Proof. Given the pullback (P, p_1, p_2) of retractions $f_i : A_i \to C$, we need to show that $(F(P), Fp_1, Fp_2)$ is a weak pullback. For that reason we shall relate it to the kernel pair of $f := [f_1, f_2] : A_1 + A_2 \to C$.

Since the f_i are retractions, i.e. right invertible, we can choose $g_i : C \to A_i$ with

$$f_i \circ g_i = id_C.$$

Let $e_i : A_i \to A_1 + A_2$ be the canonical inclusions, then

$$f \circ e_i = [f_1, f_2] \circ e_i = f_i.$$

Define $h_1 := [id_{A_1}, g_1 \circ f_2]$ and $h_2 := [g_2 \circ f_1, id_{A_2}]$, then $h_i : A_1 + A_2 \to A_i$ satisfy

$$h_i \circ e_i = id_{A_i}$$

as well as

$$f_1 \circ h_1 = [f_1 \circ id_{A_1}, f_1 \circ g_1 \circ f_2] = [f_1, f_2] = f_2 \circ h_2.$$

Thus we have established commutativity of the right half of the following diagram.

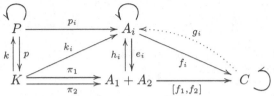

We add the kernel pair (K, π_1, π_2) of $f := [f_1, f_2]$ and the pullback (P, p_1, p_2) of f_1 and f_2. Now, the previous lemma asserts that (K, k_1, k_2) is a weak pullback of f_1 and f_2, therefore we obtain a morphism $p : P \to K$ with $k_i \circ p = p_i$.

On the other hand, (P, p_1, p_2) being the real pullback, earns us a unique morphism $k : K \to P$ with $p_i \circ k = k_i$ and $k \circ p = id_P$ by uniqueness.

Next, we apply the functor F to the above diagram. The requirements of Lemma 3 remain intact for the image diagram, and assuming that F weakly preserves the kernel (K, π_1, π_2), we obtain that $(F(K), Fk_1, Fk_2)$ is a weak pullback of Ff_1 with Ff_2. Since furthermore $F(P)$ remains being a retract of $F(K)$ by means of $Fk \circ Fp = Fid_P = id_{F(P)}$, we see that $(F(P), Fp_1, Fp_2)$, too, is a weak pullback of Ff_1 and Ff_2, as required.

From this we can obtain our mentioned joint result with Ch. Henkel. With an elementwise proof this appears in his master thesis, which has been completed under our guidance [11]:

Corollary 5. *For a Set−endofunctor F the following are equivalent:*

1. *F weakly preserves kernel pairs.*
2. *F weakly preserves pullbacks of epis.*

Proof. In *Set*, each map f can be factored as $f = \iota \circ e$ where e is epi and ι is mono. Therefore, the kernel of f is the same as the kernel of e. This takes care of the direction $(2 \to 1)$. For the other direction, the axiom of choice asserts that in *Set* epis are right invertible, so the conditions of Theorem 4 are met.

4   Free-Algebra Functors and Mal'cev Conditions

Given a finitary algebraic signature $S = (\mathfrak{f}_i, n_i)_{i \in I}$, fixing a family of function symbols $\mathfrak{f}_i$, each of arity n_i, and given a set Σ of equations, let $\mathcal{V}(\Sigma)$ be the variety defined by Σ, i.e. the class of all algebras $\mathfrak{A} = (A, (f_i^{\mathfrak{A}})_{i \in I})$ satisfying all equations from Σ.

The forgetful functor, sending an algebra $\mathfrak{A}$ from $\mathcal{V}(\Sigma)$ to its base set A, has a left adjoint F_Σ, which assigns to each set X, considered as set of variables, the free Σ-algebra over X. $F_\Sigma(X)$ consists of equivalence classes of terms p, which arise by syntactically composing basic operations named in the signature, using only variables from X.

Two terms p and q are identified if the equality $p \approx q$ is a consequence of the equations in Σ. This is the same as saying that p and q induce the same operation $p^{\mathfrak{A}} = q^{\mathfrak{A}}$ on each algebra $\mathfrak{A} \in \mathcal{V}(\Sigma)$. Instead of p we often write $p(x_1, ..., x_n)$ to mark all occurrences $x_1, ..., x_n$ of variables in the term p.

F_Σ is clearly a functor (in fact a monad), and its action on maps $\varphi : X \to Y$ can be described as variable substitution, sending $p(x_1, ..., x_n) \in F_\Sigma(X)$ to $p(\varphi x_1, ..., \varphi x_n) \in F_\Sigma(Y)$.

Σ is called *idempotent* if for every function symbol $\mathfrak{f}$ appearing in Σ we have $\Sigma \vdash \mathfrak{f}(x, ..., x) \approx x$. As a consequence, all term operations satisfy the corresponding equations, so Σ is idempotent iff $F_\Sigma(\{x\}) = \{x\}$. In this case, we also call the variety $\mathcal{V}(\Sigma)$ idempotent. As an example, the variety of all lattices is idempotent, whereas the variety of groups is not.

In [8] we have recently shown that for an idempotent set Σ of equations F_Σ weakly preserves products.

In 1954, A.I. Mal'cev [13,14] discovered that a variety has *permutable congruences*, i.e. $\Theta \circ \Psi = \Psi \circ \Theta$ holds for all congruences Θ and Ψ in each algebra of $\mathcal{V}(\Sigma)$, if and only if there exists a ternary term $m(x, y, z)$ satisfying the equations

$$x \approx m(x, y, y) \text{ and } m(x, x, y) \approx y.$$

Permutability of congruences Θ and Ψ can be generalized to *n-permutability*, requiring that the n-fold relational compositions agree:

$$\Theta \circ \Psi \circ \Theta \circ \cdots = \Psi \circ \Theta \circ \Psi \circ \cdots .$$

Here each side is meant to be the relational composition of n factors.

J. Hagemann and A. Mitschke [10], generalizing the original Mal'cev result, showed that a variety $\mathcal{V}$ is $n-$permutable if and only if there are ternary terms $p_1, ..., p_{n-1}$ such the following series of equations is satisfied:

$$x \approx p_1(x, y, y) \tag{1}$$
$$p_i(x, x, y) \approx p_{i+1}(x, y, y) \text{ for all } 0 < i < n - 1$$
$$p_{n-1}(x, x, y) \approx y.$$

Ever since Mal'cev's mentioned result, such conditions, postulating the existence of derived operations satisfying a (possibly $n-$indexed) series of equations have been called *Mal'cev conditions*, and these have played an eminent role in the development of universal algebra. They may generally be of the form

$$\exists n \in \mathbb{N}. \exists p_1, ..., p_n. \Gamma$$

where $p_1, ..., p_n$ are terms and Γ is a set of equations involving the terms $p_1, ..., p_n$. Such a condition is supposed to hold in a variety $\mathcal{V}$ of universal algebras if for some $n \in \mathbb{N}$ there exist terms $p_1, ..., p_n$ in the operations of $\mathcal{V}$ satisfying the equations in Γ.

B. Jónsson [12] gave a Mal'cev condition characterizing *congruence distributive* varieties, i.e. varieties $\mathcal{V}$ in which the lattice of congruences of each algebra

$\mathfrak{A} \in \mathcal{V}$ is distributive. A Mal'cev condition characterizing *congruence modular* varieties and involving quaternary terms was found by A. Day [2]. In 1981, employing commutator theory, we showed in [5], how to compose Jónsson's terms for congruence distributivity with the original Mal'cev term $m(x, y, z)$ from above in order to characterize congruence modular varieties, while at the same time obtaining ternary terms.

Notice, that all Mal'cev conditions mentioned above are idempotent, i.e. $\Gamma \vdash p_i(x, ..., x) \approx x$ for each of their terms p_i.

5   Preservation of Preimages

A few years ago, Dent, Kearnes, and Szendrei [3] invented a syntactic operation on idempotent sets of equations Σ, called the *derivative* Σ', and they showed that an idempotent variety $\mathcal{V}(\Sigma)$ is congruence modular if and only if $\Sigma \cup \Sigma'$ is inconsistent.

Subsequently, Freese [4] was able to give a similar characterization for n-permutable varieties, using an *order derivative*, which was based on the fact that a variety is n-permutable if and only if its algebras are not orderable, an insight, said to have been observed already by Hagemann (unpublished), but which first appears in P. Selinger's PhD-thesis [17].

It will turn out below, after generalizing the relevant definitions from [3] to non-idempotent varieties, that the derivative also serves us to characterize free-algebra functors preserving preimages.

We start by slightly modifying the definition of weak independence from [3]:

Definition 6. *A term p is* weakly independent *of a variable occurrence x, if there exists a term q such that $\Sigma \vdash p(x, v_1, ..., v_n) \approx q(y)$ where $x \neq y$ and $v_1, ..., v_n$ are variables. p is* independent *of x, if $\Sigma \models p(x, z_1, ..., z_n) \approx p(y, z_1, ..., z_n)$ where x, y, and all variables $z_1, ..., z_n$ are distinct.*

As an example, consider the the variety of groups. The term $m(x, y, z) := xy^{-1}z$ is weakly independent of its first argument, since $m(x, x, y) = xx^{-1}y \approx y$ holds, but it is not independent of the same argument, since $m(x, z_1, z_2) = xz_1^{-1}z_2 \not\approx yz_1^{-1}z_2 = m(y, z_1, z_2)$. The term $p(x, y) := xyx^{-1}$ is *independent* of its first argument in the variety of abelian groups, but not in the variety of all groups, since $p(x, z) = xzx^{-1} \not\approx yzy^{-1} = p(y, z)$. More generally, any Mal'cev term $m(x, y, z)$ is *weakly independent* of each of its arguments, but cannot be independent of either of them.

Clearly, if p is independent of x then it is also weakly independent of x, since $p(x, z_1, ..., z_n) \approx p(y, z_1, ..., z_n)$ entails $p(x, y, ..., y) \approx p(y, y, ..., y) =: q(y)$.

We now define the *derivative* Σ' *of* Σ as the set of all equations asserting that a term which is weakly independent of a variable occurrence x should also be independent of that variable:

$$\Sigma' := \{p(x, \overrightarrow{z}) \approx p(y, \overrightarrow{z}) \mid p(x, \overrightarrow{w}) \text{ weakly independent of } x\}.$$

We are now ready to state the main theorem of this section:

Theorem 7. F_Σ (weakly) preserves preimages if and only if $\Sigma \vdash \Sigma'$.

Proof. Consider a term p which is weakly independent of a variable occurrence x, i.e. there exists a term q such that $\Sigma \vdash p(x, w_1, ..., w_n) \approx q(y)$ where $x, y, w_1, ..., w_n$ are occurrences of not necessarily distinct variables, but $x \neq y$.

We must show that $p(x, z_1, ..., z_n) \approx p(y, z_1, ..., z_n)$ for $x, y, z_1, ..., z_n$ mutually distinct variables.

Consider the map $\varphi : \{x, y, z_1, ..., z_n\} \to \{x, y, w_1, ..., w_n\}$ which fixes x and y and sends each z_i to w_i. Then

$$(F_\Sigma\varphi)(p(x, z_1, ..., z_n)) = p(\varphi x, \varphi z_1, ..., \varphi z_n)$$
$$= p(x, w_1, ..., w_n)$$
$$\approx q(y) \in F_\Sigma(\{y\}),$$

so $p(x, z_1, ..., z_n)$ is in the preimage of $F_\Sigma(\{y\})$ under $F_\Sigma\varphi$.

The prcimage of $\{y\} \subseteq \{x, y, w_1, ..., w_n\}$ under φ does not contain x, so $\varphi^{-1}(\{y\}) \subseteq \{y, z_1, ..., z_n\}$. Assuming that F_Σ preserves preimages, we obtain

$$p(x, z_1, ..., z_n) \in F_\Sigma(\{y, z_1, ..., z_n\}),$$

so there exists a term r with

$$p(x, z_1, ..., z_n) \approx r(y, z_1, ..., z_n).$$

Since $x, y \notin \{z_1, ..., z_n\}$, by substituting y for x in this equation, we also find $p(y, z_1, ..., z_n) \approx r(y, z_1, ..., z_n)$, so

$$p(x, z_1, ..., z_n) \approx p(y, z_1, ..., z_n)$$

by transitivity.

Conversely, assume that $\Sigma \vdash \Sigma'$, then we need to verify that F_Σ preserves preimages. According to Lemma 2, we need only verify that F_Σ preserves classifying preimages. Let X and Y be disjoint sets and let $\varphi : X \cup Y \to \{x, y\}$ be given, sending elements from X to x and elements from Y to y. Then $Y \subseteq X \cup Y$ is the preimage of $\{y\}$ under φ. We must show that the following is a preimage diagram:

$$
\begin{array}{ccc}
F_\Sigma(X \cup Y) & \xrightarrow{F_\Sigma\varphi} & F_\Sigma(\{x, y\}) \\
\uparrow & & \uparrow \\
F_\Sigma(Y) & \longrightarrow & F_\Sigma(\{y\})
\end{array}
$$

Given an element $p \in F_\Sigma(X \cup Y)$ with $(F_\Sigma\varphi)p \in F_\Sigma(\{y\})$ we must show that $p \in F_\Sigma(Y)$.

Let $x_1, ..., x_n, y_1, ..., y_m$ be all variables occurring in p where $x_i \in X$ and $y_i \in Y$. The assumption yields

$$p(\varphi x_1, ..., \varphi x_n, \varphi y_1, ..., \varphi y_m) = p(x, ..., x, y, ..., y) \in F_\Sigma(y),$$

i.e. $p(x, ..., x, y, ..., y) \approx q(y)$ for some term $q \in F_\Sigma(\{y\})$. This means that p is weakly independent of each of its occurrences of x, so our assumption $\Sigma \vdash \Sigma'$ says that $\mathcal{V}$ also satisfies the equations

$$p(x_1, x_2, ..., x_n, y_1, ..., y_m) \approx p(z_1, x_2, ..., x_n, y_1, ..., y_m)$$
$$\approx p(z_1, z_2, ..., x_n, y_1, ..., y_m)$$
$$\approx ...$$
$$\approx p(z_1, z_2, ..., z_n, y_1, ..., y_m)$$

In particular, by choosing y arbitrarily from $\{y_1, ..., y_m\}$, we obtain

$$p = p(x_1, ..., x_n, y_1, ..., y_m) \approx p(y, ..., y, y_1, ..., y_n) \in F_\Sigma(Y)$$

as required.

Whereas Dent, Kearnes and Szendrei have shown that congruence modular varieties are characterized by the fact that adding the derivative of their defining equations yields an inconsistent theory, we have just seen that F_Σ preserving preimages delineates the other extreme, $\Sigma \vdash \Sigma'$. Clearly, therefore, for modular varieties $\mathcal{V}(\Sigma)$, the variety functor F_Σ does not preserve preimages.

6   Preservation of Kernel Pairs

We begin this section with a positive result:

Theorem 8. *If $\mathcal{V}(\Sigma)$ is a Mal'cev variety, then F_Σ weakly preserves pullbacks of epis.*

Proof. According to Corollary 5, it suffices to check that F_Σ weakly preserves kernel pairs of a surjective map $f : X \to Y$. The kernel of f is

$$K = \{(x, x') \mid f(x) = f(x')\}$$

with the cartesian projections π_1 and π_2.

Given $p := p(x_1, ..., x_m) \in F_\Sigma(X)$ and $q := q(x'_1, ..., x'_n) \in F_\Sigma(X)$, and $r := r(y_1, ..., y_k) \in F_\Sigma(Y)$ with

$$(F_\Sigma f)p = r = (F_\Sigma f)q,$$

we need to find some $\bar{m} \in F_\Sigma(K)$ such that $(F_\Sigma \pi_1)\bar{m} = p$ and $(F_\Sigma \pi_2)\bar{m} = q$.

$$F_\Sigma(K) \underset{F_\Sigma \pi_2}{\overset{F_\Sigma \pi_1}{\rightrightarrows}} F_\Sigma(X) \xrightarrow{F_\Sigma f} F_\Sigma(Y)$$

Choose g as right inverse to f, i.e. $f \circ g = id_Y$, and define:

$$\bar{p} := p((x_1, g f x_1), ..., (x_m, g f x_m)),$$
$$\bar{r} := r((g y_1, g y_1), ..., (g y_k, g y_k)),$$
$$\bar{q} := q((g f x_1', x_1'), ..., (g f x_n', x_n')),$$

then it is easy to check that $\bar{p}$, $\bar{r}$, and $\bar{q}$ are elements of $F_\Sigma(K)$, hence the same is true for $\bar{m} := m(\bar{p}, \bar{r}, \bar{q})$, where $m(x, y, z)$ is the Mal'cev term. Moreover,

$$
\begin{aligned}
(F_\Sigma \pi_1)\bar{m} &= (F_\Sigma \pi_1) m(\bar{p}, \bar{r}, \bar{q}) \\
&= m((F_\Sigma \pi_1)\bar{p}, (F_\Sigma \pi_1)\bar{r}, (F_\Sigma \pi_1)\bar{q}) \\
&= m(p(x_1, ..., x_m), r(g y_1, ..., g y_k), q(g f x_1', ..., g f x_n')) \\
&= m(p, (F_\Sigma g) r(y_1, ..., y_k), (F_\Sigma g)(F_\Sigma f) q(x_1', ..., x_n')) \\
&= m(p, (F_\Sigma g) r, (F_\Sigma g) r) \\
&= p,
\end{aligned}
$$

and similarly, $(F_\Sigma \pi_2)\bar{m} = q$.

Preservation of kernel pairs for the functor F_Σ leads to some interesting syntactic condition on terms, which might be worth of further study:

Lemma 9. *If F_Σ weakly preserves kernel pairs, then for any terms p, q we have:*

$$p(x, x, y) \approx q(x, y, y)$$

if and only if there exists a quaternary term s such that

$$p(x, y, z) \approx s(x, y, z, z)$$
$$q(x, y, z) \approx s(x, x, y, z).$$

Proof. The if-direction of the claim is obvious. For the other direction consider $\varphi, \psi : \{x, y, z\} \to \{x, z\}$ with $\varphi x = \varphi y = \psi x = x$, and $\varphi z = \psi y = \psi z = z$. Their pullback is

$$P = \{(x, x), (y, x), (z, y), (z, z)\}.$$

Since φ and ψ are epi, according to Theorem 4, their pullback should be weakly preserved, too. Now, given p and q with $p(x, x, y) \approx q(x, y, y)$, then $(F_\Sigma \varphi)p = (F_\Sigma \psi)q$. Therefore, there must be some $\cdot s \in F_\Sigma(P)$, with $(F_\Sigma \pi_1)s = p$ and $(F_\Sigma \pi_2)s = q$. We obtain:

$$p(x, y, z) \approx (F_\Sigma \pi_1)s((x, x), (y, x), (z, y), (z, z)) = s(x, y, z, z), \text{ and}$$
$$q(x, y, z) \approx (F_\Sigma \pi_2)s((x, x), (y, x), (z, y), (z, z)) = s(x, x, y, z).$$

Using the criterion of this Lemma, we now determine, which n–permutable varieties give rise to a functor weakly preserving kernel pairs:

Theorem 10. *If $\mathcal{V}(\Sigma)$ is $n-$permutable, then F_Σ weakly preserves kernel pairs if and only if $\mathcal{V}(\Sigma)$ is a Mal'cev variety, i.e. 2-permutable.*

Proof. Assuming that $\mathcal{V}(\Sigma)$ is $n-$permutable for $n > 2$, we shall show that it is already $(n-1)-$permutable.

Let $p_1, ..., p_{n-1}$ be the terms from the Mal'cev condition for n-permutability, with the Eqs. 1.

According to Lemma 9, we find some term $s(x, y, z, u)$ such that $s(x, y, z, z) \approx p_1(x, y, z)$ and $s(x, x, y, z) \approx p_2(x, y, z)$. We now define a new term

$$m(x, y, z) := s(x, y, y, z),$$

and calculate:

$$m(x, y, y) \approx s(x, y, y, y) \approx p_1(x, y, y) \approx x$$

as well as

$$m(x, x, y) \approx s(x, x, x, y) \approx p_2(x, x, y) \approx p_3(x, y, y).$$

From this we obtain a shorter chain of equations, discarding p_1 and p_2 and replacing them by m:

$$x \approx m(x, y, y)$$
$$m(x, x, y) \approx p_3(x, y, y)$$
$$p_i(x, x, y) \approx p_{i+1}(x, y, y) \text{ for all } 2 < i < n - 1$$
$$p_{n-1}(x, x, y) \approx y.$$

7   Conclusion and Further Work

We have shown that variety functors F_Σ preserve preimages if and only if $\Sigma \vdash \Sigma'$ where Σ' is the derivative of the set of equations Σ.

For each Mal'cev variety, F_Σ weakly preserves kernel pairs. For the other direction, if F_Σ weakly preserves kernel pairs, then every equation of the shape $p(x, x, y) \approx q(x, y, y)$ ensures the existence of a term $s(x, y, z, u)$, such that $p(x, y, z) \approx s(x, y, z, z)$ and $q(x, y, y) \approx s(x, x, y, z)$.

This intriguing algebraic condition appears to be new and certainly deserves further study from a universal algebraic perspective. Adding it to Freese's characterization of n-permutable varieties by means of his "order-derivative", allows to distinguish between the cases $n = 2$ (Mal'cev varieties) and $n > 2$, which cannot be achieved using his order derivative, alone.

It would be desirable to see if this consequence of weak kernel preservation can be turned into a concise if-and-only-if statement.

References

1. Barr, M.: Terminal coalgebras in well-founded set theory. Theoret. Comput. Sci. **114**(2), 299–315 (1993). https://doi.org/10.1016/0304-3975(93)90076-6
2. Day, A.: A characterization of modularity for congruence lattices of algebras. Can. Math. Bull. **12**(2), 167–173 (1969). https://doi.org/10.4153/CMB-1969-016-6
3. Dent, T., Kearnes, K.A., Szendrei, Á.: An easy test for congruence modularity. Algebra Univers. **67**(4), 375–392 (2012). https://doi.org/10.1007/s00012-012-0186-z
4. Freese, R.: Equations implying congruence n-permutability and semidistributivity. Algebra Univers. **70**(4), 347–357 (2013). https://doi.org/10.1007/s00012-013-0256-x
5. Gumm, H.P.: Congruence modularity is permutability composed with distributivity. Arch. Math **36**, 569–576 (1981). https://doi.org/10.1007/BF01223741
6. Gumm, H.P.: Elements of the general theory of coalgebras. In: LUATCS 99. Rand Afrikaans University, Johannesburg, South Africa (1999). https://www.mathematik.uni-marburg.de/~gumm/Papers/Luatcs.pdf
7. Fiadeiro, J.L., Harman, N., Roggenbach, M., Rutten, J. (eds.): CALCO 2005. LNCS, vol. 3629. Springer, Heidelberg (2005). https://doi.org/10.1007/11548133
8. Gumm, H.P.: Connected monads weakly preserve products. Algebra Univers. **81**(2), 18 (2020). https://doi.org/10.1007/s00012-020-00654-w
9. Gumm, H.P., Schröder, T.: Types and coalgebraic structure. Algebra Univers. **53**, 229–252 (2005). https://doi.org/10.1007/s00012-005-1888-2
10. Hagemann, J., Mitschke, A.: On n-permutable congruences. Algebra Univers. **3**(1), 8–12 (1973). https://doi.org/10.1007/BF02945100
11. Henkel, C.: Klassifikation Coalgebraischer Typfunktoren. Diplomarbeit, Universität Marburg (2010)
12. Jónsson, B.: Algebras whose congruence lattices are distributive. Math. Scand. **21**, 110–121 (1967). https://doi.org/10.7146/math.scand.a-10850
13. Mal'cev, A.I.: On the general theory of algebraic systems. Matematicheskii Sbornik (N.S.) **35**(77), 3–20 (1954). (in Russian)
14. Mal'tsev, A.I.: On the general theory of algebraic systems. Am. Math. Soc. **27**, 125–142 (1963). Transl., Ser. 2
15. Rutten, J.: Universal coalgebra: a theory of systems. Theoret. Comput. Sci. **249**, 3–80 (2000). https://doi.org/10.1016/S0304-3975(00)00056-6
16. Rutten, J.: Universal coalgebra: a theory of systems. Techical report, CWI, Amsterdam (1996). CS-R9652
17. Selinger, P.: Functionality, polymorphism, and concurrency: a mathematical investigation of programming paradigms. Technical report, IRCS 97–17, University of Pennsylvania (1997). https://repository.upenn.edu/ircs_reports/88/
18. Trnková, V.: On descriptive classification of set-functors. I. Comm. Math. Univ. Carolinae **12**, 143–174 (1971). https://eudml.org/doc/16419

Learning Automata with Side-Effects

Gerco van Heerdt[1]([✉])[iD], Matteo Sammartino[1,2][iD], and Alexandra Silva[1][iD]

[1] University College London, London, UK
{gerco.heerdt,alexandra.silva}@ucl.ac.uk
[2] Royal Holloway, University of London, London, UK
matteo.sammartino@rhul.ac.uk

Abstract. Automata learning has been successfully applied in the verification of hardware and software. The size of the automaton model learned is a bottleneck for scalability, and hence optimizations that enable learning of compact representations are important. This paper exploits monads, both as a mathematical structure and a programming construct, to design and prove correct a wide class of such optimizations. Monads enable the development of a new learning algorithm and correctness proofs, building upon a general framework for automata learning based on category theory. The new algorithm is parametric on a monad, which provides a rich algebraic structure to capture non-determinism and other side-effects. We show that this allows us to uniformly capture existing algorithms, develop new ones, and add optimizations.

Keywords: Automata · Learning · Side-effects · Monads · Algebras

1 Introduction

The increasing complexity of software and hardware systems calls for new scalable methods to design, verify, and continuously improve systems. Black-box inference methods aim at building models of running systems by observing their response to certain queries. This reverse engineering process is very amenable for automation and allows for fine-tuning the precision of the model depending on the properties of interest, which is important for scalability.

One of the most successful instances of black-box inference is automata learning, which has been used in various verification tasks, ranging from finding bugs in implementations of network protocols [15] to rejuvenating legacy software [29]. Vaandrager [30] has written a comprehensive overview of the widespread use of automata learning in verification.

A limitation in automata learning is that the models of real systems can become too large to be handled by tools. This demands compositional methods and techniques that enable compact representation of behaviors.

This work was partially supported by the ERC Starting Grant ProFoundNet (grant code 679127), a Leverhulme Prize (PLP–2016–129), and the EPSRC Standard Grant CLeVer (EP/S028641/1).

In this paper, we show how monads can be used to add optimizations to learning algorithms in order to obtain compact representations. We will use as playground for our approach the well known L* algorithm [2], which learns a minimal deterministic finite automaton (DFA) accepting a regular language by interacting with a *teacher*, i.e., an oracle that can reply to specific queries about the target language. Monads allow us to take an abstract approach, in which category theory is used to devise an optimized learning algorithm and a generic correctness proof for a broad class of compact models.

The inspiration for this work is quite concrete: it is a well-known fact that non-deterministic finite automata (NFAs) can be much smaller than deterministic ones for a regular language. The subtle point is that given a regular language, there is a canonical deterministic automaton accepting it—the minimal one—but there might be many "minimal" non-deterministic automata accepting the same language. This raises a challenge for learning algorithms: which non-deterministic automaton should the algorithm learn? To overcome this, Bollig et al. [11] developed a version of Angluin's L* algorithm, which they called NL*, in which they use a particular class of NFAs, *Residual Finite State Automata* (RFSAs), which do admit minimal canonical representatives. Though NL* indeed is a first step in incorporating a more compact representation of regular languages, there are several questions that remain to be addressed. We tackle them in this paper.

DFAs and NFAs are formally connected by the subset construction. Underpinning this construction is the rich algebraic structure of languages and of the state space of the DFA obtained by determinizing an NFA. The state space of a determinized DFA—consisting of subsets of the state space of the original NFA—has a join-semilattice structure. Moreover, this structure is preserved in language acceptance: if there are subsets U and V, then the language of $U \cup V$ is the union of the languages of the first two. Formally, the function that assigns to each state its language is a join-semilattice map, since languages themselves are just sets of words and have a lattice structure. And languages are even richer: they have the structure of complete atomic Boolean algebras. This leads to several questions: Can we exploit this structure and have even more compact representations? What if we slightly change the setting and look at weighted languages over a semiring, which have the structure of a semimodule (or vector space, if the semiring is a field)?

The latter question is strongly motivated by the widespread use of weighted languages and corresponding *weighted finite automata* (WFAs) in verification, from the formal verification of quantitative properties [13,17,25], to probabilistic model-checking [5], to the verification of on-line algorithms [1].

Our key insight is that the algebraic structures mentioned above are in fact algebras for a monad T. In the case of join-semilattices this is the powerset monad, and in the case of vector spaces it is the free vector space monad. These monads can be used to define a notion of T-automaton, with states having the structure of an algebra for the monad T, which generalizes non-determinism as a side-effect. From T-automata we can derive a compact, equivalent version by taking as states a set of *generators* and transferring the algebraic structure of the original state space to the transition structure of the automaton.

This general perspective enables us to generalize L^* to a new algorithm L^*_T, which learns compact automata featuring non-determinism and other side-effects captured by a monad. Moreover, L^*_T incorporates further optimizations arising from the monadic representation, which lead to more scalable algorithms.

We start by giving an overview of our approach, which also states our main contributions in greater detail and ends with a road map of the rest of the paper.

2   Overview and Contributions

In this section, we explain the original L^* algorithm and discuss the challenges in adapting the algorithm to learn automata with side-effects, illustrating them through a concrete example—NFAs. We then highlight our main contributions.

L^* Algorithm. This algorithm learns the minimal DFA accepting a language $\mathcal{L} \subseteq A^*$ over a finite alphabet A. It assumes the existence of a *minimally adequate teacher*, which is an oracle that can answer two types of queries: 1. *Membership queries*: given a word $w \in A^*$, does w belong to $\mathcal{L}$? and 2. *Equivalence queries*: given a *hypothesis* DFA $\mathcal{H}$, does $\mathcal{H}$ accept $\mathcal{L}$? If not, the teacher returns a *counterexample*, i.e., a word incorrectly classified by $\mathcal{H}$. The algorithm incrementally builds an *observation table* made of two parts: a top part, with rows ranging over a finite set $S \subseteq A^*$; and a bottom part, with rows ranging over $S \cdot A$ ($\cdot$ is pointwise concatenation). Columns range over a finite $E \subseteq A^*$. For each $u \in S \cup S \cdot A$ and $v \in E$, the corresponding cell in the table contains 1 if and only if $uv \in \mathcal{L}$. Intuitively, each row u contains enough information to fully identify the Myhill–Nerode equivalence class of u with respect to an approximation of the target language—rows with the same content are considered members of the same equivalence class. Cells are filled in using membership queries.

As an example, and to set notation, consider the table below over $A = \{a, b\}$. It shows that $\mathcal{L}$ contains the word aa and does not contain the words ε (the empty word), a, b, ba, aaa, and baa.

$$
\begin{array}{c}
 \overbrace{}^{E} \\
\begin{array}{cc|ccc}
 & & \varepsilon & a & aa \\
\hline
S\,[& \varepsilon & 0 & 0 & 1 \\
\hline
S \cdot A\,[& a & 0 & 1 & 0 \\
 & b & 0 & 0 & 0
\end{array}
\end{array}
\qquad
\begin{array}{l}
\mathsf{row_t} : S \to 2^E \qquad \mathsf{row_t}(u)(v) = 1 \iff uv \in \mathcal{L} \\[1em]
\mathsf{row_b} : S \to (2^E)^A \quad \mathsf{row_b}(u)(a)(v) = 1 \iff uav \in \mathcal{L}
\end{array}
$$

We use functions $\mathsf{row_t}$ and $\mathsf{row_b}$ to describe the top and bottom parts of the table, respectively. Notice that S and $S \cdot A$ may intersect. For conciseness, when tables are depicted, elements in the intersection are only shown in the top part.

A key idea of the algorithm is to construct a hypothesis DFA from the different rows in the table. The construction is the same as that of the minimal DFA from the Myhill–Nerode equivalence, and exploits the correspondence between table rows and Myhill–Nerode equivalence classes. The state space of the hypothesis DFA is given by the set $H = \{\mathsf{row_t}(s) \mid s \in S\}$. Note that there may

```
1   S, E ← {ε}
2   repeat
3       while the table is not closed or not consistent
4           if the table is not closed
5               find t ∈ S, a ∈ A such that row_b(t)(a) ≠ row_t(s) for all s ∈ S
6                   S ← S ∪ {ta}
7           if the table is not consistent
8               find s_1, s_2 ∈ S, a ∈ A, and e ∈ E such that
                    row_t(s_1) = row_t(s_2) and row_b(s_1)(a)(e) ≠ row_b(s_2)(a)(e)
9                   E ← E ∪ {ae}
10      Construct the hypothesis H and submit it to the teacher
11      if the teacher replies no, with a counterexample z
12          S ← S ∪ prefixes(z)
13  until the teacher replies yes
14  return H
```

Fig. 1. L* algorithm.

be multiple rows with the same content, but they result in a single state, as they all belong to the same Myhill–Nerode equivalence class. The initial state is $\text{row}_t(\varepsilon)$, and we use the ε column to determine whether a state is accepting: $\text{row}_t(s)$ is accepting whenever $\text{row}_t(s)(\varepsilon) = 1$. The transition function is defined as $\text{row}_t(s) \xrightarrow{a} \text{row}_b(s)(a)$. (Notice that the continuation is drawn from the bottom part of the table). For the hypothesis automaton to be well-defined, ε must be in S and E, and the table must satisfy two properties:

- **Closedness** states that each transition actually leads to a state of the hypothesis. That is, the table is closed if for all $t \in S$ and $a \in A$ there is $s \in S$ such that $\text{row}_t(s) = \text{row}_b(t)(a)$.
- **Consistency** states that there is no ambiguity in determining the transitions. That is, the table is consistent if for all $s_1, s_2 \in S$ such that $\text{row}_t(s_1) = \text{row}_t(s_2)$ we have $\text{row}_b(s_1) = \text{row}_b(s_2)$.

The algorithm updates the sets S and E to satisfy these properties, constructs a hypothesis, submits it in an equivalence query, and, when given a counterexample, refines the hypothesis. This process continues until the hypothesis is correct. The algorithm is shown in Fig. 1.

Example Run. We now run the algorithm with the target language $\mathcal{L} = \{w \in \{a\}^* \mid |w| \neq 1\}$. The minimal DFA accepting $\mathcal{L}$ is

$$\mathcal{M} = \quad \to \bigcirc \xrightarrow{\ a\ } \bigcirc \xrightarrow{\ a\ } \bigcirc\!\!\circlearrowright a \tag{1}$$

Initially, $S = E = \{\varepsilon\}$. We build the observation table given in Fig. 2a. This table is not closed, because the row with label a, having 0 in the only column, does not appear in the top part of the table: the only row ε has 1. To fix this,

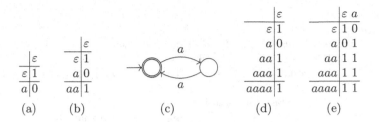

Fig. 2. Example run of L* on $\mathcal{L} = \{w \in \{a\}^* \mid |w| \neq 1\}$.

we add the word a to the set S. Now the table (Fig. 2b) is closed and consistent, so we construct the hypothesis that is shown in Fig. 2c and pose an equivalence query. The teacher replies *no* and informs us that the word aaa should have been accepted. L* handles a counterexample by adding all its prefixes to the set S. We only have to add aa and aaa in this case. The next table (Fig. 2d) is closed, but not consistent: the rows ε and aa both have value 1, but their extensions a and aaa differ. To fix this, we prepend the continuation a to the column ε on which they differ and add $a \cdot \varepsilon = a$ to E. This distinguishes $\mathsf{row}_t(\varepsilon)$ from $\mathsf{row}_t(aa)$, as seen in the next table in Fig. 2e. The table is now closed and consistent, and the new hypothesis automaton is precisely the correct one $\mathcal{M}$.

As mentioned, the hypothesis construction approximates the theoretical construction of the minimal DFA, which is unique up to isomorphism. That is, for $S = E = A^*$ the relation that identifies words of S having the same value in row_t is precisely the Myhill–Nerode's right congruence.

Learning Non-deterministic Automata. As is well known, NFAs can be smaller than the minimal DFA for a given language. For example, the language $\mathcal{L}$ above is accepted by the NFA

$$\mathcal{N} = \quad \text{(2)}$$

which is smaller than the minimal DFA $\mathcal{M}$. Though in this example, which we chose for simplicity, the state reduction is not massive, it is known that in general NFAs can be exponentially smaller than the minimal DFA [24]. This reduction of the state space is enabled by a side-effect—non-determinism, in this case.

Learning NFAs can lead to a substantial gain in space complexity, but it is challenging. The main difficulty is that NFAs do not have a canonical minimal representative: there may be several non-isomorphic state-minimal NFAs accepting the same language, which poses problems for the development of the learning algorithm. To overcome this, Bollig et al. [11] proposed to use a particular class of NFAs, namely RFSAs, which do admit minimal canonical representatives. However, their ad-hoc solution for NFAs does not extend to other automata, such as weighted or alternating. In this paper we present a solution that works for any side-effect, specified as a monad.

The crucial observation underlying our approach is that the language semantics of an NFA is defined in terms of its determinization, i.e., the DFA obtained by taking sets of states of the NFA as state space. In other words, this DFA is defined over an algebraic structure induced by the powerset, namely a (complete) *join semilattice* (JSL) whose join operation is set union. This automaton model does admit minimal representatives, which leads to the key idea for our algorithm: learning NFAs as automata over JSLs. In order to do so, we use an extended table where rows have a JSL structure, defined as follows. The join of two rows is given by an element-wise or, and the bottom element is the row containing only zeroes. More precisely, the new table consists of the two functions

$$\mathsf{row}_t^\sharp\colon \mathcal{P}(S) \to 2^E \qquad \mathsf{row}_b^\sharp\colon \mathcal{P}(S) \to (2^E)^A$$

given by $\mathsf{row}_t^\sharp(U) = \bigvee\{\mathsf{row}_t(s) \mid s \in U\}$ and $\mathsf{row}_b^\sharp(U)(a) = \bigvee\{\mathsf{row}_b(s)(a) \mid s \in U\}$. Formally, these functions are JSL homomorphisms, and they induce the following general definitions:

- The table is *closed* if for all $U \subseteq S, a \in A$ there is $U' \subseteq S$ such that $\mathsf{row}_t^\sharp(U') = \mathsf{row}_b^\sharp(U)(a)$.
- The table is *consistent* if for all $U_1, U_2 \subseteq S$ s.t. $\mathsf{row}_t^\sharp(U_1) = \mathsf{row}_t^\sharp(U_2)$ we have $\mathsf{row}_b^\sharp(U_1) = \mathsf{row}_b^\sharp(U_2)$.

We remark that our algorithm does not actually store the whole extended table, which can be quite large. It only needs to store the original table over S because all other rows in $\mathcal{P}(S)$ are freely generated and can be computed as needed, with no additional membership queries. The only lines in Fig. 1 that need to be adjusted are lines 5 and 8, where closedness and consistency are replaced with the new notions given above. Moreover, $\mathcal{H}$ is now built from the extended table.

Optimizations. In this paper we also present two optimizations to our algorithm. For the first one, note that the state space of the hypothesis constructed by the algorithm can be very large since it encodes the entire algebraic structure. We show that we can extract a *minimal set of generators* from the table and compute a *succinct hypothesis* in the form of an automaton with side-effects, without any algebraic structure. For JSLs, this consists in only taking rows that are not the join of other rows, i.e., the *join-irreducibles*. By applying this optimization to this specific case, we essentially recover the learning algorithm of Bollig et al. [11]. The second optimization is a generalization of the optimized counterexample handling method of Rivest and Schapire [28], originally intended for L* and DFAs. It consists in processing counterexamples by adding a single *suffix* of the counterexample to E, instead of adding all prefixes of the counterexample to S. This can avoid the algorithm posing a large number of membership queries.

Example Revisited. We now run the new algorithm on the language $\mathcal{L} = \{w \in \{a\}^* \mid |w| \neq 1\}$ considered earlier. Starting from $S = E = \{\varepsilon\}$, the observation table (Fig. 3a) is immediately closed and consistent. (It is closed because we

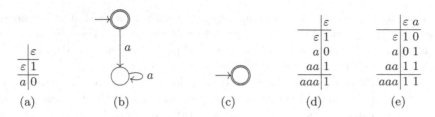

Fig. 3. Example run of the L^* adaptation for NFAs on $\mathcal{L} = \{w \in \{a\}^* \mid |w| \neq 1\}$.

have $\mathsf{row}_t^\sharp(\{a\}) = \mathsf{row}_t^\sharp(\emptyset)$.) This gives the JSL hypothesis shown in Fig. 3b, which leads to an NFA hypothesis having a single state that is initial, accepting, and has no transitions (Fig. 3c). The hypothesis is incorrect, and the teacher may supply us with counterexample aa. Adding prefixes a and aa to S leads to the table in Fig. 3d. The table is again closed, but not consistent: $\mathsf{row}_t^\sharp(\{a\}) = \mathsf{row}_t^\sharp(\emptyset)$, but $\mathsf{row}_b^\sharp(\{a\})(a) = \mathsf{row}_t^\sharp(\{aa\}) \neq \mathsf{row}_t^\sharp(\emptyset) = \mathsf{row}_b^\sharp(\emptyset)(a)$. Thus, we add a to E. The resulting table (Fig. 3e) is closed and consistent. We note that row aa is the union of other rows: $\mathsf{row}_t^\sharp(\{aa\}) = \mathsf{row}_t^\sharp(\{\varepsilon, a\})$ (i.e., it is not a join-irreducible), and therefore can be ignored when building the succinct hypothesis. This hypothesis has two states, ε and a, and indeed it is the correct one $\mathcal{N}$.

Contributions and Road Map of the Paper. After some preliminary notions in Sect. 3, we present the main contributions:

– In Sect. 4, we develop a general algorithm L_T^*, which generalizes the NFA one presented in Sect. 2 to an arbitrary *monad* T capturing side-effects, and we provide a general correctness proof for our algorithm.
– In Sect. 5, we describe the first optimization and prove its correctness.
– In Sect. 6 we describe the second optimization. We also show how it can be combined with the one of Sect. 5, and how it can lead to a further small optimization, where the consistency check on the table is dropped.
– Finally, in Sect. 7 we show how L_T^* can be applied to several automata models, highlighting further case-specific optimizations when available.

3   Preliminaries

In this section we define a notion of T-automaton, a generalization of non-deterministic finite automata parametric in a monad T. We assume familiarity with basic notions of category theory: functors (in the category **Set** of sets and functions) and natural transformations.

Side-effects can be conveniently captured as a *monad*. A monad $T = (T, \eta, \mu)$ is a triple consisting of an endofunctor T on **Set** and two natural transformations: a *unit* $\eta \colon \mathrm{Id} \Rightarrow T$ and a *multiplication* $\mu \colon T^2 \Rightarrow T$, which satisfy the compatibility laws $\mu \circ \eta_T = \mathrm{id}_T = \mu \circ T\eta$ and $\mu \circ \mu_T = \mu \circ T\mu$.

Example 1 (Monads). An example of a monad is the triple $(\mathcal{P}, \{-\}, \bigcup)$, where $\mathcal{P}$ denotes the powerset functor associating a collection of subsets to a set, $\{-\}$

is the singleton operation, and $\bigcup$ is just union of sets. Another example is the triple $(V(-), e, m)$, where $V(X)$ is the free semimodule (over a semiring $\mathbb{S}$) over X, namely $\{\varphi \mid \varphi \colon X \to \mathbb{S}$ having finite support$\}$. The support of a function $\varphi \colon X \to \mathbb{S}$ is the set of $x \in X$ such that $\varphi(x) \neq 0$. Then $e \colon X \to V(X)$ is the characteristic function for each $x \in X$, and $m \colon V(V(X)) \to V(X)$ is defined for $\varphi \in V(V(X))$ and $x \in X$ as $m(\varphi)(x) = \sum_{\psi \in V(X)} \varphi(\psi) \times \psi(x)$.

Given a monad T, a T-algebra is a pair (X, h) consisting of a carrier set X and a function $h \colon TX \to X$ such that $h \circ \mu_X = h \circ Th$ and $h \circ \eta_X = \mathrm{id}_X$. A T-homomorphism between two T-algebras (X, h) and (Y, k) is a function $f \colon X \to Y$ such that $f \circ h = k \circ Tf$. The abstract notion of T-algebra instantiates to expected notions, as illustrated in the following example.

Example 2 (Algebras for a monad). The $\mathcal{P}$-algebras are the (complete) join-semilattices, and their homomorphisms are join-preserving functions. If $\mathbb{S}$ is a field, V-algebras are vector spaces, and their homomorphisms are linear maps.

We will often refer to a T-algebra (X, h) as X if h is understood or if its specific definition is irrelevant. Given a set X, (TX, μ_X) is a T-algebra called the *free T-algebra* on X. One can build algebras pointwise for some operations. For instance, if Y is a set and (X, x) a T-algebra, then we have a T-algebra (X^Y, f), where $f \colon T(X^Y) \to X^Y$ is given by $f(W)(y) = (x \circ T(\mathsf{ev}_y))(W)$ and $\mathsf{ev}_y \colon X^Y \to X$ by $\mathsf{ev}_y(g) = g(y)$. If U and V are T-algebras and $f \colon U \to V$ is a T-algebra homomorphism, then the image $\mathrm{img}(f)$ of f is a T-algebra, with the T-algebra structure inherited from V. The following proposition connects algebra homomorphisms from the free T-algebra on a set U to an algebra V with functions $U \to V$. We will make use of this later in the section.

Proposition 3. *Given a set U and a T-algebra (V, v), there is a bijective correspondence between T-algebra homomorphisms $TU \to V$ and functions $U \to V$: for a T-algebra homomorphism $f \colon TU \to V$, define $f^\dagger = f \circ \eta \colon U \to V$; for a function $g \colon U \to V$, define $g^\sharp = v \circ Tg \colon TU \to V$. Then $g^\sharp$ is a T-algebra homomorphism called the* free T-extension *of g, and we have $f^{\dagger\sharp} = f$ and $g^{\sharp\dagger} = g$.*

We now have all the ingredients to define our notion of automaton with side-effects and their language semantics. We fix a monad (T, η, μ) with T preserving finite sets, as well as a T-algebra O that models outputs of automata.

Definition 4 (T-automaton). *A T-automaton is a quadruple $(Q, \delta \colon Q \to Q^A, \mathsf{out} \colon Q \to O, \mathsf{init} \in Q)$, where Q is a T-algebra, the transition map δ and output map out are T-algebra homomorphisms, and init is the initial state.*

Example 5. DFAs are Id-automata when $O = 2 = \{0, 1\}$ is used to distinguish accepting from rejecting states. For the more general case of O being any set, DFAs generalize into *Moore automata*.

Example 6. Recall that $\mathcal{P}$-algebras are JSLs, and their homomorphisms are join-preserving functions. In a $\mathcal{P}$-automaton, Q is equipped with a join operation, and

Q^A is a join-semilattice with pointwise join: $(f \vee g)(a) = f(a) \vee g(a)$ for $a \in A$. Since the automaton maps preserve joins, we have, in particular, $\delta(q_1 \vee q_2)(a) = \delta(q_1)(a) \vee \delta(q_2)(a)$. One can represent an NFA over a set of states S as a $\mathcal{P}$-automaton by taking $Q = (\mathcal{P}(S), \bigcup)$ and $O = 2$, the Boolean join-semilattice with the *or* operation as its join. Let init $\subseteq S$ be the set of initial states and out: $\mathcal{P}(Q) \rightarrow 2$ and $\delta \colon \mathcal{P}(S) \rightarrow \mathcal{P}(S)^A$ the respective extensions (Proposition 3) of the NFA's output and transition functions. The resulting $\mathcal{P}$-automaton is precisely the determinized version of the NFA.

More generally, an automaton with side-effects given by a monad T always represents a T-automaton with a free state space.

Proposition 7. *A T-automaton of the form $((TX, \mu_X), \delta, \mathsf{out}, \mathsf{init})$, for any set X, is completely defined by the set X with the element $\mathsf{init} \in TX$ and functions*

$$\delta^\dagger \colon X \rightarrow (TX)^A \qquad \mathsf{out}^\dagger \colon X \rightarrow O.$$

We call such a T-automaton a *succinct* automaton, which we sometimes identify with the representation $(X, \delta^\dagger, \mathsf{out}^\dagger, \mathsf{init})$. These automata are closely related to the ones studied in [18].

A *(generalized) language* is a function $\mathcal{L} \colon A^* \rightarrow O$. For every T-automaton we have an *observability* and a *reachability* map, telling respectively which state is reached by reading a given word and which language each state recognizes.

Definition 8 (Reachability/Observability Maps). *The* reachability map *of a T-automaton $\mathcal{A}$ is a function $r_\mathcal{A} \colon A^* \rightarrow Q$ inductively defined as: $r_\mathcal{A}(\varepsilon) = \mathsf{init}$ and $r_\mathcal{A}(ua) = \delta(r_\mathcal{A}(u))(a)$. The* observability map *of $\mathcal{A}$ is a function $o_\mathcal{A} \colon Q \rightarrow O^{A^*}$ given by: $o_\mathcal{A}(q)(\varepsilon) = \mathsf{out}(q)$ and $o_\mathcal{A}(q)(av) = o_\mathcal{A}(\delta(q)(a))(v)$.*

The *language accepted by* $\mathcal{A}$ is the map $\mathcal{L}_\mathcal{A} = o_\mathcal{A}(\mathsf{init}) = \mathsf{out}_\mathcal{A} \circ r_\mathcal{A} \colon A^* \rightarrow O$.

Example 9. For an NFA $\mathcal{A}$ represented as a $\mathcal{P}$-automaton, as seen in Example 6, $o_\mathcal{A}(q)$ is the language of q in the traditional sense. Note that q, in general, is a set of states: $o_\mathcal{A}(q)$ takes the union of languages of singleton states. The set $\mathcal{L}_\mathcal{A}$ is the language accepted by the initial states, i.e., the language of the NFA. The reachability map $r_\mathcal{A}(u)$ returns the set of states reached via all paths reading u.

Given a language $\mathcal{L} \colon A^* \rightarrow O$, there exists a (unique) *minimal T-automaton* $\mathcal{M}_\mathcal{L}$ accepting $\mathcal{L}$, which is minimal in the number of states. Its existence follows from general facts. See for example [19].

Definition 10 (Minimal T-Automaton for $\mathcal{L}$). *Let $t_\mathcal{L} \colon A^* \rightarrow O^{A^*}$ be the function giving the residual languages of $\mathcal{L}$, namely $t_\mathcal{L}(u) = \lambda v.\mathcal{L}(uv)$. The minimal T-automaton $\mathcal{M}_\mathcal{L}$ accepting $\mathcal{L}$ has state space $M = \mathsf{img}(t_\mathcal{L}^\sharp)$, initial state $\mathsf{init} = t_\mathcal{L}(\varepsilon)$, and T-algebra homomorphisms $\mathsf{out} \colon M \rightarrow O$ and $\delta \colon M \rightarrow M^A$ given by $\mathsf{out}(t_\mathcal{L}^\sharp(U)) = \mathcal{L}(U)$ and $\delta(t_\mathcal{L}^\sharp(U))(a)(v) = t_\mathcal{L}^\sharp(U)(av)$.*

In the following, we will also make use of the *minimal Moore automaton* accepting $\mathcal{L}$. Although this always exists—by instantiating Definition 10 with $T = \mathsf{Id}$—it need not be finite. The following property says that finiteness of Moore automata and of T-automata accepting the same language are related.

Proposition 11. *The minimal Moore automaton accepting $\mathcal{L}$ is finite if and only if the minimal T-automaton accepting $\mathcal{L}$ is finite.*

4   A General Algorithm

In this section we introduce our extension of L^* to learn automata with side-effects. The algorithm is parametric in the notion of side-effect, represented as the monad T, and is therefore called L_T^*. We fix a language $\mathcal{L}\colon A^* \to O$ that is to be learned, and we assume that there is a finite T-automaton accepting $\mathcal{L}$. This assumption generalizes the requirement of L^* that $\mathcal{L}$ is regular (i.e., accepted by a specific class of T-automata, see Example 5).

An observation table consists of a pair of functions

$$\mathrm{row}_t\colon S \to O^E \qquad\qquad \mathrm{row}_b\colon S \to (O^E)^A$$

given by $\mathrm{row}_t(s)(e) = \mathcal{L}(se)$ and $\mathrm{row}_b(s)(a)(e) = \mathcal{L}(sae)$, where $S, E \subseteq A^*$ are finite sets with $\varepsilon \in S \cap E$. For $O = 2$, we recover exactly the L^* observation table. The key idea for L_T^* is defining closedness and consistency over the free T-extensions of those functions.

Definition 12 (Closedness and Consistency). *The table is* closed *if for all $U \in T(S)$ and $a \in A$ there exists a $U' \in T(S)$ such that $\mathrm{row}_t^\sharp(U') = \mathrm{row}_b^\sharp(U)(a)$. The table is* consistent *if for all $U_1, U_2 \in T(S)$ such that $\mathrm{row}_t^\sharp(U_1) = \mathrm{row}_t^\sharp(U_2)$ we have $\mathrm{row}_b^\sharp(U_1) = \mathrm{row}_b^\sharp(U_2)$.*

For closedness, we do not need to check all elements of $T(S) \times A$ against elements of $T(S)$, but only those of $S \times A$, thanks to the following result.

Lemma 13. *If for all $s \in S$ and $a \in A$ there is $U \in T(S)$ such that $\mathrm{row}_t^\sharp(U) = \mathrm{row}_b(s)(a)$, then the table is closed.*

Example 14. For NFAs represented as $\mathcal{P}$-automata, the properties are as presented in Sect. 2. Recall that for $T = \mathcal{P}$ and $O = 2$, the Boolean join-semilattice, $\mathrm{row}_t^\sharp$ and $\mathrm{row}_b^\sharp$ describe a table where rows are labeled by subsets of S. Then we have, for instance, $\mathrm{row}_t^\sharp(\{s_1, s_2\})(e) = \mathrm{row}_t(s_1)(e) \vee \mathrm{row}_t(s_2)(e)$, i.e., $\mathrm{row}_t^\sharp(\{s_1, s_2\})(e) = 1$ if and only if $\mathcal{L}(s_1 e) = 1$ or $\mathcal{L}(s_2 e) = 1$. Closedness amounts to check whether each row in the bottom part of the table is the join of a set of rows in the top part. Consistency amounts to check whether, for all sets of rows $U_1, U_2 \subseteq S$ in the top part of the table whose joins are equal, the joins of rows $U_1 \cdot \{a\}$ and $U_2 \cdot \{a\}$ in the bottom part are also equal, for all $a \in A$.

If closedness and consistency hold, we can define a *hypothesis* T-automaton $\mathcal{H}$, with state space $H = \mathrm{img}(\mathrm{row}_t^\sharp)$, $\mathrm{init} = \mathrm{row}_t(\varepsilon)$, and output and transitions

$$\mathrm{out}\colon H \to O \qquad\qquad \mathrm{out}(\mathrm{row}_t^\sharp(U)) = \mathrm{row}_t^\sharp(U)(\varepsilon)$$
$$\delta\colon H \to H^A \qquad\qquad \delta(\mathrm{row}_t^\sharp(U)) = \mathrm{row}_b^\sharp(U).$$

```
1  S, E ← {ε}
2  repeat
3       while the table is not closed or not consistent
4            if the table is not closed
5                 find s ∈ S, a ∈ A such that rowb(s)(a) ≠ rowt♯(U) for all U ∈ T(S)
6                 S ← S ∪ {sa}
7            if the table is not consistent
8                 find U1, U2 ∈ T(S), a ∈ A, and e ∈ E such that
                      rowt♯(U1) = rowt♯(U2) and rowb♯(U1)(a)(e) ≠ rowb♯(U2)(a)(e)
9                 E ← E ∪ {ae}
10      Construct the hypothesis ℋ and submit it to the teacher
11      if the teacher replies no, with a counterexample z
12           S ← S ∪ prefixes(z)
13 until the teacher replies yes
14 return ℋ
```

Fig. 4. Adaptation of L^* for T-automata.

The correctness of this definition follows from the abstract treatment of [21], instantiated to the category of T-algebras and their homomorphisms.

We can now give algorithm L^*_T. Similarly to the example in Sect. 2, we only have to adjust lines 5 and 8 in Fig. 1. The resulting algorithm is shown in Fig. 4.

Correctness. Correctness for L^*_T amounts to proving that, for any target language $\mathcal{L}$, the algorithm terminates returning the minimal T-automaton $\mathcal{M}_\mathcal{L}$ accepting $\mathcal{L}$. As in the original L^* algorithm, we only need to prove that the algorithm terminates, that is, that only finitely many hypotheses are produced. Correctness follows from termination, since line 13 causes the algorithm to terminate only if the hypothesis automaton coincides with $\mathcal{M}_\mathcal{L}$.

In order to show termination, we argue that the state space H of the hypothesis increases while the algorithm loops, and that H cannot be larger than M, the state space of $\mathcal{M}_\mathcal{L}$. In fact, when a closedness defect is resolved (line 6), a row that was not previously found in the image of $\mathsf{row}_t^\sharp \colon T(S) \to O^E$ is added, so the set H grows larger. When a consistency defect is resolved (line 9), two previously equal rows become distinguished, which also increases the size of H.

As for counterexamples, adding their prefixes to S (line 11) creates a consistency defect, which will be fixed during the next iteration, causing H to increase. This is due to the following result, which says that the counterexample z has a prefix that violates consistency. Note that the hypothesis $\mathcal{H}$ in the statement below is the hypothesis obtained before adding the prefixes of z to S.

Proposition 15. *If* $z \in A^*$ *is such that* $\mathcal{L}_\mathcal{H}(z) \neq \mathcal{L}(z)$ *and* $\mathsf{prefixes}(z) \subseteq S$, *then there are a prefix* ua *of* z, *with* $u \in A^*$ *and* $a \in A$, *and* $U \in T(S)$ *such that* $\mathsf{row}_t(u) = \mathsf{row}_t^\sharp(U)$ *and* $\mathsf{row}_b(u)(a) \neq \mathsf{row}_b^\sharp(U)(a)$.

Now, note that by increasing S or E, the hypothesis state space H never decreases in size. Moreover, for $S = A^*$ and $E = A^*$, $\mathsf{row}_t^\sharp = t_\mathcal{L}^\sharp$. Therefore, since H and M are defined as the images of $\mathrm{row}_t^\sharp$ and $t_L^\sharp$, respectively, the size of H is bounded by that of M. Since H increases while the algorithm loops, the algorithm must terminate and is thus correct.

Note that the learning algorithm of Bollig et al. does not terminate using this counterexample processing method [10, Appendix F]. This is due to their notion of consistency being weaker than ours: we have shown that progress is guaranteed because a consistency defect, in our sense, is created using this method.

Query Complexity. The complexity of automata learning algorithms is usually measured in terms of the number of both membership and equivalence queries asked, as it is common to assume that computations within the algorithm are insignificant compared to evaluating the system under analysis in applications. The cost of answering queries themselves is not considered, as it depends on the implementation of the teacher, which the algorithm abstracts from.

The table is a T-algebra homomorphism, so membership queries for rows labeled in S are enough to determine all other rows. We measure the query complexities in terms of the number of states n of the minimal Moore automaton, the number of states t of the minimal T-automaton, the size k of the alphabet, and the length m of the longest counterexample. Note that t cannot be smaller than n, but it can be much bigger. For example, when $T = \mathcal{P}$, t may be in $\mathcal{O}(2^n)$.[1]

The maximum number of closedness defects fixed by the algorithm is n, as a closedness defect for the setting with algebraic structure is also a closedness defect for the setting without that structure. The maximum number of consistency defects fixed by the algorithm is t, as fixing a consistency defect distinguishes two rows that were previously identified. Since counterexamples lead to consistency defects, this also means that the algorithm will not pose more than t equivalence queries. A word is added to S when fixing a closedness defect, and $\mathcal{O}(m)$ words are added to S when processing a counterexample. The number of rows that we need to fill using queries is therefore in $\mathcal{O}(tmk)$. The number of columns added to the table is given by the number of times a consistency defect is fixed and thus in $\mathcal{O}(t)$. Altogether, the number of membership queries is in $\mathcal{O}(t^2mk)$.

5   Succinct Hypotheses

We now describe the first of two optimizations, which is enabled by the use of monads. Our algorithm produces hypotheses that can be quite large, as their state space is the image of $\mathrm{row}_t^\sharp$, which has the whole set $T(S)$ as its domain. For instance, when $T = \mathcal{P}$, $T(S)$ is exponentially larger than S. We show how we can compute *succinct* hypotheses, whose state space is given by a subset of S. We start by defining sets of *generators for the table*.

---

[1] Take the language $\{a^p\}$, for some $p \in \mathbb{N}$ and a singleton alphabet $\{a\}$. Its residual languages are $\emptyset$ and $\{a^i\}$ for all $0 \le i \le p$, thus the minimal DFA accepting the language has $p+2$ states. However, the residual languages w.r.t. sets of words are all the subsets of $\{\varepsilon, a, aa, \dots, a^p\}$—hence, the minimal T-automaton has 2^{p+1} states.

Definition 16. *A set $S' \subseteq S$ is a set of generators for the table whenever for all $s \in S$ there is $U \in T(S')$ such that $\mathsf{row_t}(s) = \mathsf{row}_t^{\sharp}(U)$.*[2]

Intuitively, U is the decomposition of s into a "combination" of generators. When $T = \mathcal{P}$, S' generates the table whenever each row can be obtained as the join of a set of rows labeled by S'. Explicitly: for all $s \in S$ there is $\{s_1, \ldots, s_n\} \subseteq S'$ such that $\mathsf{row_t}(s) = \mathsf{row}_t^{\sharp}(\{s_1, \ldots, s_n\}) = \mathsf{row_t}(s_1) \vee \cdots \vee \mathsf{row_t}(s_n)$.

Recall that $\mathcal{H}$, with state space H, is the hypothesis automaton for the table. The existence of generators S' allows us to compute a T-automaton with state space $T(S')$ equivalent to $\mathcal{H}$. We call this the *succinct hypothesis*, although $T(S')$ may be larger than H. Proposition 7 tells us that the succinct hypothesis can be represented as an automaton with side-effects in T that has S' as its state space. This results in a lower space complexity when storing the hypothesis.

We now show how the succinct hypothesis is computed. Observe that, if generators S' exist, $\mathsf{row}_t^{\sharp}$ factors through the restriction of itself to $T(S')$. Denote this latter function $\widehat{\mathsf{row}_t}^{\sharp}$. Since we have $T(S') \subseteq T(S)$, the image of $\widehat{\mathsf{row}_t}^{\sharp}$ coincides with $\mathsf{img}(\mathsf{row}_t^{\sharp}) = H$, and therefore the surjection restricting $\widehat{\mathsf{row}_t}^{\sharp}$ to its image has the form $e \colon T(S') \to H$. Any right inverse $i \colon H \to T(S')$ of the function e (that is, $e \circ i = \mathsf{id}_H$, but whereas e is a T-algebra homomorphism, i need not be one) yields a succinct hypothesis as follows.

Definition 17 (Succinct Hypothesis). *The succinct hypothesis is the T-automaton $\mathcal{S} = (T(S'), \delta, \mathsf{out}, \mathsf{init})$ given by $\mathsf{init} = i(\mathsf{row_t}(\varepsilon))$ and*

$$\mathsf{out}^{\dagger} \colon S' \to O \qquad\qquad \mathsf{out}^{\dagger}(s) = \mathsf{row_t}(s)(\varepsilon)$$
$$\delta^{\dagger} \colon S' \to T(S')^A \qquad\qquad \delta^{\dagger}(s)(a) = i(\mathsf{row_b}(s)(a)).$$

This definition is inspired by that of a *scoop*, due to Arbib and Manes [4].

Proposition 18. *Any succinct hypothesis of $\mathcal{H}$ accepts the language of $\mathcal{H}$.*

We now give a simple procedure to compute a *minimal* set of generators, that is, a set S' such that no proper subset is a set of generators. This generalizes a procedure defined by Angluin et al. [3] for non-deterministic, universal, and alternating automata.

Proposition 19. *The following algorithm returns a minimal set of generators for the table:*

$S' \leftarrow S$
while *there are $s \in S'$ and $U \in T(S' \setminus \{s\})$ s.t. $\mathsf{row}_t^{\sharp}(U) = \mathsf{row_t}(s)$*
$\qquad S' \leftarrow S' \setminus \{s\}$
return S'

---

[2] Here and hereafter we assume that $T(S') \subseteq T(S)$, and more generally that T preserves inclusion maps. To eliminate this assumption, one could take the inclusion map $f \colon S' \hookrightarrow S$ and write $\mathsf{row}_t^{\sharp}(T(f)(U))$ instead of $\mathsf{row}_t^{\sharp}(U)$.

To determine whether U as in the above algorithm exists, one can always naively enumerate all possibilities, using that T preserves finite sets. This is what we call the basic algorithm. For specific algebraic structures, one may find more efficient methods, as we show in the following example.

Example 20. Consider the powerset monad $T = \mathcal{P}$. We now exemplify two ways of computing succinct hypotheses, which are inspired by canonical RFSAs [16]. The basic idea is to start from a deterministic automaton and to remove states that are equivalent to a set of other states. The algorithm given in Proposition 19 computes a minimal S' that only contains labels of rows that are not the join of other rows. (In case two rows are equal, only one of their labels is kept.) In other words, as mentioned in Sect. 2, S' contains labels of join-irreducible rows. To concretize the algorithm efficiently, we use a method introduced by Bollig et al. [11], which essentially exploits the natural order on the JSL of table rows. In contrast to the basic exponential algorithm, this results in a polynomial one.[3] Bollig et al. determine whether a row is a join of other rows by comparing the row just to the join of rows below it. Like them, we make use of this also to compute right inverses of e, for which we will formalize the order.

The function $e \colon \mathcal{P}(S') \to H$ tells us which sets of rows are equivalent to a single state in H. We show two right inverses $H \to \mathcal{P}(S')$ for it. The first one,

$$i_1(h) = \{s \in S' \mid \mathrm{row}_t(s) \leq h\},$$

stems from the construction of the *canonical RFSA* of a language [16]. Here we use the order $a \leq b \iff a \vee b = b$ induced by the JSL structure. The resulting construction of a succinct hypothesis was first used by Bollig et al. [11]. This succinct hypothesis has a "maximal" transition function, meaning that no more transitions can be added without changing the language of the automaton.

The second inverse is

$$i_2(h) = \{s \in S' \mid \mathrm{row}_t(s) \leq h \text{ and for all } s' \in S' \text{ s.t. } \mathrm{row}_t(s) \leq \mathrm{row}_t(s') \leq h$$
$$\text{we have } \mathrm{row}_t(s) = \mathrm{row}_t(s')\},$$

resulting in a more economical transition function, where some redundancies are removed. This corresponds to the *simplified canonical RFSA* [16].

Example 21. Consider $T = \mathcal{P}$, and recall the table in Fig. 3e. When $S' = S$, the right inverse given by i_1 yields the succinct hypothesis shown below.

Note that $i_1(\mathrm{row}_t(aa)) = \{\varepsilon, a, aa\}$. Taking i_2 instead, the succinct hypothesis is just the DFA (1) because $i_2(\mathrm{row}_t(aa)) = \{aa\}$. Rather than constructing a

---

[3] When we refer to computational complexities, as opposed to query complexities, they are in terms of the sizes of S, E, and A.

succinct hypothesis directly, our algorithm first reduces the set S'. In this case, we have $\text{row}_t(aa) = \text{row}_t^\sharp(\{\varepsilon, a\})$, so we remove aa from S'. Now i_1 and i_2 coincide and produce the NFA (2). Minimizing the set S' in this setting essentially comes down to determining what Bollig et al. [11] call the *prime* rows of the table.

Remark 22. The algorithm in Proposition 19 implicitly assumes an order in which elements of S are checked. Although the algorithm is correct for any such order, different orders may give results that differ in size.

6   Optimized Counterexample Handling

The second optimization we give generalizes the counterexample processing method due to Rivest and Schapire [28], which improves the worst case complexity of the number of membership queries needed in L*. Maler and Pnueli [26] proposed to add all suffixes of the counterexample to the set E instead of adding all prefixes to the set S. This eliminates the need for consistency checks in the deterministic setting. The method by Rivest and Schapire finds a *single* suffix of the counterexample and adds it to E. This suffix is chosen in such a way that it either distinguishes two existing rows or creates a closedness defect, both of which imply that the hypothesis automaton will grow.

The main idea is finding the distinguishing suffix via the hypothesis automaton $\mathcal{H}$. Given $u \in A^*$, let q_u be the state in $\mathcal{H}$ reached by reading u, i.e., $q_u = r_{\mathcal{H}}(u)$. For each $q \in H$, we pick any $U_q \in T(S)$ that yields q according to the table, i.e., such that $\text{row}_t^\sharp(U_q) = q$. Then for a counterexample z we have that the residual language w.r.t. U_{q_z} does not "agree" with the residual language w.r.t. z.

The above intuition can be formalized as follows. Let $\mathcal{R}\colon A^* \to O^{A^*}$ be given by $\mathcal{R}(u) = t_{\mathcal{L}}^\sharp(U_{q_u})$ for all $u \in A^*$, the residual language computation. We have the following technical lemma, saying that a counterexample z distinguishes the residual languages $t_{\mathcal{L}}(z)$ and $\mathcal{R}(z)$.

Lemma 23. *If $z \in A^*$ is such that $\mathcal{L}_{\mathcal{H}}(z) \neq \mathcal{L}(z)$, then $t_{\mathcal{L}}(z)(\varepsilon) \neq \mathcal{R}(z)(\varepsilon)$.*

We assume that $U_{q_\varepsilon} = \eta(\varepsilon)$. For a counterexample z, we then have $\mathcal{R}(\varepsilon)(z) = t_{\mathcal{L}}(\varepsilon)(z) \neq \mathcal{R}(z)(\varepsilon)$. While reading z, the hypothesis automaton passes a sequence of states $q_{u_0}, q_{u_1}, q_{u_2}, \ldots, q_{u_n}$, where $u_0 = \epsilon$, $u_n = z$, and $u_{i+1} = u_i a$ for some $a \in A$ is a prefix of z. If z were correctly classified by $\mathcal{H}$, all residuals $\mathcal{R}(u_i)$ would classify the remaining suffix v of z, i.e., such that $z = u_i v$, in the same way. However, the previous lemma tells us that, for a counterexample z, this is not case, meaning that for some suffix v we have $\mathcal{R}(ua)(v) \neq \mathcal{R}(u)(av)$. In short, this inequality is discovered along a transition in the path to z.

Corollary 24. *If $z \in A^*$ is such that $\mathcal{L}_{\mathcal{H}}(z) \neq \mathcal{L}(z)$, then there are $u, v \in A^*$ and $a \in A$ such that $uav = z$ and $\mathcal{R}(ua)(v) \neq \mathcal{R}(u)(av)$.*

To find such a decomposition efficiently, Rivest and Schapire use a binary search algorithm. We conclude with the following result that turns the above property into the elimination of a closedness witness. That is, given a counterexample z and the resulting decomposition uav from the above corollary, we show that, while currently $\mathrm{row}_t^\sharp(U_{q_{ua}}) = \mathrm{row}_b^\sharp(U_{q_u})(a)$, after adding v to E we have $\mathrm{row}_t^\sharp(U_{q_{ua}})(v) \neq \mathrm{row}_b^\sharp(U_{q_u})(a)(v)$. (To see that the latter follows from the proposition below, note that for all $U \in T(S)$ and $e \in E$, $\mathrm{row}_t^\sharp(U)(e) = t_{\mathcal{L}}^\sharp(U)(e)$ and for each $a' \in A$, $\mathrm{row}_b^\sharp(U)(a')(e) = t_{\mathcal{L}}^\sharp(U)(a'e)$.) The inequality means that either we have a closedness defect, or there still exists some $U \in T(S)$ such that $\mathrm{row}_t^\sharp(U) = \mathrm{row}_b^\sharp(U_{q_u})(a)$. In this case, the rows $\mathrm{row}_t^\sharp(U)$ and $\mathrm{row}_t^\sharp(U_{q_{ua}})$ have become distinguished by adding v, which means that the size of H has increased. A closedness defect also increases the size of H, so in any case we make progress.

Proposition 25. *If $z \in A^*$ is such that $\mathcal{L}_{\mathcal{H}}(z) \neq \mathcal{L}(z)$, then there are $u, v \in A^*$ and $a \in A$ such that $\mathrm{row}_t^\sharp(U_{q_{ua}}) = \mathrm{row}_b^\sharp(U_{q_u})(a)$ and $t_{\mathcal{L}}^\sharp(U_{q_{ua}})(v) \neq t_{\mathcal{L}}^\sharp(U_{q_u})(av)$.*

We now show how to combine this optimized counterexample processing method with the succinct hypothesis optimization from Sect. 5. Recall that the succinct hypothesis S is based on a right inverse $i \colon H \to T(S')$ of $e \colon T(S') \to H$. Choosing such an i is equivalent to choosing U_q for each $q \in H$. We then redefine $\mathcal{R}$ using the reachability map of the succinct hypothesis. Specifically, $\mathcal{R}(u) = t_{\mathcal{L}}^\sharp(r_S(u))$ for all $u \in A^*$.

Unfortunately, there is one complication. We assumed earlier that $U_{q_\varepsilon} = \eta(\varepsilon)$, or more specifically $\mathcal{R}(\varepsilon)(z) = \mathcal{L}(z)$. This now may be impossible because we do not even necessarily have $\varepsilon \in S'$. We show next that if this equality does not hold, then there are two rows that we can distinguish by adding z to E. Thus, after testing whether $\mathcal{R}(\varepsilon)(z) = \mathcal{L}(z)$, we either add z to E (if the test fails) or proceed with the original method.

Proposition 26. *If $z \in A^*$ is such that $\mathcal{R}(\varepsilon)(z) \neq \mathcal{L}(z)$, then $\mathrm{row}_t^\sharp(\mathrm{init}_S) = \mathrm{row}_t(\varepsilon)$ and $t_{\mathcal{L}}^\sharp(\mathrm{init}_S)(z) \neq t_{\mathcal{L}}(\varepsilon)(z)$.*

To see that the original method still works, we prove the analogue of Proposition 25 for the new definition of $\mathcal{R}$.

Proposition 27. *If $z \in A^*$ is such that $\mathcal{L}_S(z) \neq \mathcal{L}(z)$ and $\mathcal{R}(\varepsilon)(z) = \mathcal{L}(z)$, then there are $u, v \in A^*$ and $a \in A$ such that $\mathrm{row}_t^\sharp(r_S^\dagger(ua)) = \mathrm{row}_b^\sharp(r_S^\dagger(u))(a)$ and $t_{\mathcal{L}}^\sharp(r_S^\dagger(ua))(v) \neq t_{\mathcal{L}}^\sharp(r_S^\dagger(u))(av)$.*

Example 28. Recall the succinct hypothesis S from Fig. 3c for the table in Fig. 2a. Note that $S' = S$ cannot be further reduced. The hypothesis is based on the right inverse $i \colon H \to \mathcal{P}(S)$ of $e \colon \mathcal{P}(S) \to H$ given by $i(\mathrm{row}_t(\varepsilon)) = \{\varepsilon\}$ and $i(\mathrm{row}_t^\sharp(\emptyset)) = \emptyset$. This is the only possible right inverse because e is bijective. For the prefixes of the counterexample aa we have $r_S(\varepsilon) = \{\varepsilon\}$ and $r_S(a) = r_S(aa) = \emptyset$. Note that $t_{\mathcal{L}}^\sharp(\{\varepsilon\})(aa) = 1$ while $t_{\mathcal{L}}(\emptyset)(a) = t_{\mathcal{L}}(\emptyset)(\varepsilon) = 0$. Thus, $\mathcal{R}(\varepsilon)(aa) \neq \mathcal{R}(a)(a)$. Adding a to E would indeed create a closedness defect.

Query Complexity. Again, we measure the membership and equivalence query complexities in terms of the number of states n of the minimal Moore automaton, the number of states t of the minimal T-automaton, the size k of the alphabet, and the length m of the longest counterexample.

A counterexample now gives an additional column instead of a set of rows, and we have seen that this leads to either a closedness defect or to two rows being distinguished. Thus, the number of equivalence queries is still at most t, and the number of columns is still in $\mathcal{O}(t)$. However, the number of rows that we need to fill using membership queries is now in $\mathcal{O}(nk)$. This means that a total of $\mathcal{O}(tnk)$ membership queries is needed to fill the table.

Apart from filling the table, we also need queries to analyze counterexamples. The binary search algorithm mentioned after Corollary 24 requires for each counterexample $\mathcal{O}(\log m)$ computations of $\mathcal{R}(x)(y)$ for varying words x and y. Let r be the maximum number of queries required for a single such computation. Note that for $u, v \in A^*$, and letting $\alpha \colon TO \to O$ be the algebra structure on O, we have $\mathcal{R}(u)(v) = \alpha(T(\mathsf{ev}_v \circ t_{\mathcal{L}})(U_{q_u}))$ for the original definition of $\mathcal{R}$ and

$$\mathcal{R}(u)(v) = \alpha(T(\mathsf{ev}_v \circ t_{\mathcal{L}})(r_{\mathcal{S}}^{\dagger}(u)))$$

in the succinct hypothesis case. Since the restricted map $T(\mathsf{ev}_v \circ t_{\mathcal{L}}) \colon TS \to TO$ is completely determined by $\mathsf{ev}_v \circ t_{\mathcal{L}} \colon S \to O$, r is at most $|S|$, which is bounded by n in this optimized algorithm. For some examples (see for instance the writer automata in Sect. 7), we even have $r = 1$. The overall membership query complexity is $\mathcal{O}(tnk + tr \log m)$.

Dropping Consistency. We described the counterexample processing method based around Proposition 25 in terms of the succinct hypothesis $\mathcal{S}$ rather than the actual hypothesis $\mathcal{H}$ by showing that $\mathcal{R}$ can be defined using $\mathcal{S}$. Since the definition of the succinct hypothesis does not rely on the property of consistency to be well-defined, this means we could drop the consistency check from the algorithm altogether. We can still measure progress in terms of the size of the set H, but it will not be the state space of an actual hypothesis during intermediate stages. This observation also explains why Bollig et al. [11] are able to use a weaker notion of consistency in their algorithm. Interestingly, they exploit the canonicity of their choice of succinct hypotheses to arrive at a polynomial membership query complexity that does not involve the factor t.

7   Examples

In this section we list several examples that can be seen as T-automata and hence learned via an instance of $\mathsf{L}_T^\star$. We remark that, since our algorithm operates on finite structures (recall that T preserves finite sets), for each automaton type one can obtain a basic, correct-by-construction instance of $\mathsf{L}_T^\star$ for free, by plugging the concrete definition of the monad into the abstract algorithm. However, we note that this is not how $\mathsf{L}_T^\star$ is intended to be used in a real-world context; it should be seen as an abstract specification of the operations each concrete implementation needs to perform, or, in other words, as a template for real implementations.

For each instance below, we discuss whether certain operations admit a more efficient implementation than the basic one, based on the specific algebraic structure induced by the monad. Due to our general treatment, the optimizations of Sects. 5 and 6 apply to all of these instances.

Non-deterministic Automata. As discussed before, non-deterministic automata are $\mathcal{P}$-automata with a free state space, provided that $O = 2$ is equipped with the "or" operation as its $\mathcal{P}$-algebra structure. We also mentioned that, as Bollig et al. [11] showed, there is a polynomial time algorithm to check whether a given row is the join of other rows. This gives an efficient method for handling closedness straight away. Moreover, as shown in Example 20, it allows for an efficient construction of the succinct hypothesis. Unfortunately, checking for consistency defects seems to require a number of computations exponential in the number of rows. However, as explained at the end of Sect. 6, we can in fact drop consistency altogether.

Universal Automata. Just like non-deterministic automata, universal automata can be seen as $\mathcal{P}$-automata with a free state space. The difference is that the $\mathcal{P}$-algebra structure on $O = 2$ is dual: it is given by the "and" rather than the "or" operation. Universal automata accept a word when all paths reading that word are accepting. One can dualize the optimized specific algorithms for the case of non-deterministic automata. This is precisely what Angluin et al. [3] have done.

Partial Automata. Consider the *maybe monad* $\mathtt{Maybe}(X) = 1 + X$, with natural transformations having components $\eta_X : X \to 1 + X$ and $\mu_X : 1 + 1 + X \to 1 + X$ defined in the standard way. Partial automata with states X can be represented as $\mathtt{Maybe}$-automata with state space $\mathtt{Maybe}(X) = 1 + X$, where there is an additional *sink state*, and output algebra $O = \mathtt{Maybe}(1) = 1 + 1$. Here the left value is for rejecting states, including the sink one. The transition map $\delta : 1 + X \to (1 + X)^A$ represents an undefined transition as one going to the sink state. The algorithm $\mathrm{L}^*_{\mathtt{Maybe}}$ is mostly like L^*, except that implicitly the table has an additional row with zeroes in every column. Since the monad only adds a single element to each set, there is no need to optimize the basic algorithm for this specific case.

Weighted Automata. Recall from Sect. 3 the *free semimodule monad* V, sending a set X to the free semimodule over a finite semiring $\mathbb{S}$. Weighted automata over a set of states X can be represented as V-automata whose state space is the semimodule $V(X)$, the output function out: $V(X) \to \mathbb{S}$ assigns a weight to each state, and the transition map $\delta : V(X) \to V(X)^A$ sends each state and each input symbol to a linear combination of states. The obvious semimodule structure on $\mathbb{S}$ extends to a pointwise structure on the potential rows of the table. The basic algorithm loops over all linear combinations of rows to check closedness and over all pairs of combinations of rows to check consistency, making them extremely expensive operations. If $\mathbb{S}$ is a field, a row can be decomposed into a linear combination of other rows in polynomial time using standard techniques from linear algebra. As a result, there are efficient procedures for checking closedness and constructing succinct hypotheses. It was shown by Van Heerdt et al. [21] that consistency in this setting is equivalent to closedness of the transpose of the table. This trick is due to Bergadano and Varricchio [7], who first studied learning of weighted automata.

Alternating Automata. We use the characterization of alternating automata due to Bertrand and Rot [9]. Recall that, given a partially ordered set $(P, \leq)$, an *upset* is a subset U of P such that, if $x \in U$ and $x \leq y$, then $y \in U$. Given $Q \subseteq P$, we write $\uparrow Q$ for the *upward closure* of Q, that is the smallest upset of P containing Q. We consider the monad A that maps a set X to the set of all upsets of $\mathcal{P}(X)$. Its unit is given by $\eta_X(x) = \uparrow\{\{x\}\}$ and its multiplication by

$$\mu_X(U) = \{V \subseteq X \mid \exists_{W \in U} \forall_{Y \in W} \exists_{Z \in Y} Z \subseteq V\}.$$

Algebras for the monad A are *completely distributive lattices* [27]. The sets of sets in $A(X)$ can be seen as DNF formulae over elements of X, where the outer powerset is disjunctive and the inner one is conjunctive. Accordingly, we define an algebra structure $\beta \colon A(2) \to 2$ on the output set 2 by letting $\beta(U) = 1$ if $\{1\} \in U$, 0 otherwise. Alternating automata with states X can be represented as A-automata with state space $A(X)$, output map $\text{out} \colon A(X) \to 2$, and transition map $\delta \colon A(X) \to A(X)^A$, sending each state to a DNF formula over X. The only difference with the usual definition of alternating automata is that $A(X)$ is not the full set $\mathcal{P}\mathcal{P}(X)$, which is not a monad [23]. However, for each formula in $\mathcal{P}\mathcal{P}(X)$ there is an equivalent one in $A(X)$.

An adaptation of L^* for alternating automata was introduced by Angluin et al. [3] and further investigated by Berndt et al. [8]. The former found that given a row $r \in 2^E$ and a set of rows $X \subseteq 2^E$, r is equal to a DNF combination of rows from X (where logical operators are applied component-wise) if and only if it is equal to the combination defined by $Y = \{\{x \in X \mid x(e) = 1\} \mid e \in E \wedge r(e) = 1\}$. We can reuse this idea to efficiently find closedness defects and to construct the hypothesis. Even though the monad A formally requires the use of DNF formulae representing upsets, in the actual implementation we can use smaller formulae, e.g., Y above instead of its upward closure. In fact, it is easy to check that DNF combinations of rows are invariant under upward closure. Similar as before, we do not know of an efficient way to ensure consistency, but we could drop it.

Writer Automata. The examples considered so far involve existing classes of automata. To further demonstrate the generality of our approach, we introduce a new (as far as we know) type of automaton, which we call *writer automaton*.

The *writer monad* $\text{Writer}(X) = \mathbb{M} \times X$ for a finite monoid M has a unit $\eta_X \colon X \to \mathbb{M} \times X$ given by adding the unit e of the monoid, $\eta_X(x) = (e, x)$, and a multiplication $\mu_X \colon \mathbb{M} \times \mathbb{M} \times X \to \mathbb{M} \times X$ given by performing the monoid multiplication, $\mu_X(m_1, m_2, x) = (m_1 m_2, x)$. In Haskell, the writer monad is used for such tasks as collecting successive log messages, where the monoid is given by the set of sets or lists of possible messages and the multiplication adds a message.

The algebras for this monad are sets Q equipped with an $\mathbb{M}$-action. One may take the output object to be the set $\mathbb{M}$ with the monoid multiplication as its action. Writer-automata with a free state space can be represented as deterministic automata that have an element of $\mathbb{M}$ associated with each transition. The semantics is as expected: $\mathbb{M}$-elements multiply along paths and finally multiply with the output of the last state to produce the actual output.

The basic learning algorithm has polynomial time complexity. To determine whether a given row is a combination of rows in the table, i.e., whether it is given by a monoid value applied to one of the rows in the table, one simply tries all of these values. This allows us to check for closedness, to minimize the generators, and to construct the succinct hypothesis, in polynominal time. Consistency involves comparing all ways of applying monoid values to rows and, for each comparison, at most $|A|$ further comparisons between one-letter extensions. The total number of comparisons is clearly polynomial in $|\mathbb{M}|$, $|S|$, and $|A|$.

8   Conclusion

We have presented L_T^*, a general adaptation of L^* that uses monads to learn an automaton with algebraic structure, as well as a method for finding a succinct equivalent based on its generators. Furthermore, we adapted the optimized counterexample handling method of Rivest and Schapire [28] to this setting and discussed instantiations to non-deterministic, universal, partial, weighted, alternating, and writer automata.

Related Work. This paper builds on and extends the theoretical toolkit of Van Heerdt et al. [19,21], who are developing a categorical automata learning framework (CALF) in which learning algorithms can be understood and developed in a structured way.

An adaptation of L^* that produces NFAs was first developed by Bollig et al. [11]. Their algorithm learns a special subclass of NFAs consisting of RFSAs, which were introduced by Denis et al. [16]. Anguin et al. [3] unified algorithms for NFAs, universal automata, and alternating automata, the latter of which was further improved by Berndt et al. [8]. We are able to provide a more general framework, which encompasses and goes beyond those classes of automata. Moreover, we study optimized counterexample handling, which [3,8,11] do not consider.

The algorithm for weighted automata over an arbitrary field was studied in a category theoretical context by Jacobs and Silva [22] and elaborated on by Van Heerdt et al. [21]. The algorithm itself was introduced by Bergadano and Varricchio [7]. The theory of succinct automata used for our hypotheses is based on the work of Arbib and Manes [4], revamped to more recent category theory.

Future Work. Whereas our general algorithm effortlessly instantiates to monads that preserve finite sets, a major challenge lies in investigating monads that do not enjoy this property. The algorithm for weighted automata generalizes to an infinite field [7,21,22] and even a principal ideal domain [20]. However, for an infinite semiring in general we cannot guarantee termination, which is because a finitely generated semimodule may have an infinite chain of strict submodules [20]. Intuitively, this means that while fixing closedness defects increases the size of the hypothesis state space semimodule, an infinite number of steps may be needed to resolve all closedness defects. In future work we would like to characterize more precisely for which semirings we can learn, and ideally formulate this characterization on the monad level.

As a result of the correspondence between learning and conformance testing [6,21], it should be possible to include in our framework the W-method [14], which is often used in case studies deploying L* (e.g. [12,15]). We defer a thorough investigation of conformance testing to future work.

References

1. Aminof, B., Kupferman, O., Lampert, R.: Reasoning about online algorithms with weighted automata. ACM Trans. Algorithms **6**(2), 28:1–28:36 (2010)
2. Angluin, D.: Learning regular sets from queries and counterexamples. Inf. Comput. **75**, 87–106 (1987)
3. Angluin, D., Eisenstat, S., Fisman, D.: Learning regular languages via alternating automata. In: IJCAI, pp. 3308–3314 (2015)
4. Arbib, M.A., Manes, E.G.: Fuzzy machines in a category. Bull. AMS **13**, 169–210 (1975)
5. Baier, C., Größer, M., Ciesinski, F.: Model checking linear-time properties of probabilistic systems. In: Droste, M., Kuich, W., Vogler, H. (eds.) Handbook of Weighted Automata. EATCS, pp. 519–570. Springer, Heidelberg (2009)
6. Berg, T., Grinchtein, O., Jonsson, B., Leucker, M., Raffelt, H., Steffen, B.: On the correspondence between conformance testing and regular Inference. In: Cerioli, M. (ed.) FASE 2005. LNCS, vol. 3442, pp. 175–189. Springer, Heidelberg (2005). https://doi.org/10.1007/978-3-540-31984-9_14
7. Bergadano, F., Varricchio, S.: Learning behaviors of automata from multiplicity and equivalence queries. SIAM J. Comput. **25**, 1268–1280 (1996)
8. Berndt, S., Liśkiewicz, M., Lutter, M., Reischuk, R.: Learning residual alternating automata. In: AAAI, pp. 1749–1755 (2017)
9. Bertrand, M., Rot, J.: Coalgebraic determinization of alternating automata. arXiv preprint arXiv:1804.02546 (2018)
10. Bollig, B., Habermehl, P., Kern, C., Leucker, M.: Angluin-style learning of NFA (research report LSV-08-28). Technical report, ENS Cachan (2008)
11. Bollig, B., Habermehl, P., Kern, C., Leucker, M.: Angluin-style learning of NFA. In: IJCAI, vol. 9, pp. 1004–1009 (2009)
12. Chalupar, G., Peherstorfer, S., Poll, E., de Ruiter, J.: Automated reverse engineering using Lego®. In: WOOT (2014)
13. Chatterjee, K., Doyen, L., Henzinger, T.A.: Quantitative languages. In: CSL, pp. 385–400 (2008)

14. Chow, T.S.: Testing software design modeled by finite-state machines. IEEE Trans. Softw. Eng. **4**, 178–187 (1978)
15. de Ruiter, J., Poll, E.: Protocol state fuzzing of TLS implementations. In: USENIX Security, pp. 193–206 (2015)
16. Denis, F., Lemay, A., Terlutte, A.: Residual finite state automata. Fundamenta Informaticae **51**, 339–368 (2002)
17. Droste, M., Gastin, P.: Weighted automata and weighted logics. In: Caires, L., Italiano, G.F., Monteiro, L., Palamidessi, C., Yung, M. (eds.) ICALP 2005. LNCS, vol. 3580, pp. 513–525. Springer, Heidelberg (2005). https://doi.org/10.1007/11523468_42
18. Goncharov, S., Milius, S., Silva, A.: Towards a coalgebraic Chomsky hierarchy. In: Diaz, J., Lanese, I., Sangiorgi, D. (eds.) TCS 2014. LNCS, vol. 8705, pp. 265–280. Springer, Heidelberg (2014). https://doi.org/10.1007/978-3-662-44602-7_21
19. van Heerdt, G.: An abstract automata learning framework. Master's thesis, Radboud University Nijmegen (2016)
20. van Heerdt, G., Kupke, C., Rot, J., Silva, A.: Learning weighted automata over principal ideal domains. arXiv preprint arXiv:1911.04404 (2019)
21. van Heerdt, G., Sammartino, M., Silva, A.: CALF: categorical automata learning framework. In: CSL, pp. 29:1–29:24 (2017)
22. Jacobs, B., Silva, A.: Automata learning: a categorical perspective. In: van Breugel, F., Kashefi, E., Palamidessi, C., Rutten, J. (eds.) Horizons of the Mind. A Tribute to Prakash Panangaden. LNCS, vol. 8464, pp. 384–406. Springer, Cham (2014). https://doi.org/10.1007/978-3-319-06880-0_20
23. Klin, B., Salamanca, J.: Iterated covariant powerset is not a monad. Electron. Notes Theor. Comput. Sci. **341**, 261–276 (2018)
24. Kozen, D.C.: Automata and Computability. Springer, New York (2012)
25. Kuperberg, D.: Linear temporal logic for regular cost functions. Log. Methods Comput. Sci. **10**(1) (2014)
26. Maler, O., Pnueli, A.: On the learnability of infinitary regular sets. Inf. Comput. **118**, 316–326 (1995)
27. Markowsky, G.: Free completely distributive lattices. Proc. Am. Math. Soc. **74**(2), 227–228 (1979)
28. Rivest, R.L., Schapire, R.E.: Inference of finite automata using homing sequences. Inf. Comput. **103**, 299–347 (1993)
29. Schuts, M., Hooman, J., Vaandrager, F.: Refactoring of legacy software using model learning and equivalence checking: an industrial experience report. In: Ábrahám, E., Huisman, M. (eds.) IFM 2016. LNCS, vol. 9681, pp. 311–325. Springer, Cham (2016). https://doi.org/10.1007/978-3-319-33693-0_20
30. Vaandrager, F.W.: Model learning. Commun. ACM **60**(2), 86–95 (2017)

De Finetti's Construction
as a Categorical Limit

Bart Jacobs[1(✉)] and Sam Staton[2]

[1] Institute for Computing and Information Sciences,
Radboud University, Nijmegen, NL, The Netherlands
bart@cs.ru.nl
[2] Department of Computer Science, University of Oxford, Oxford, UK
sam.staton@cs.ox.ac.uk

Abstract. This paper reformulates a classical result in probability theory from the 1930s in modern categorical terms: de Finetti's representation theorem is redescribed as limit statement for a chain of finite spaces in the Kleisli category of the Giry monad. This new limit is used to identify among exchangeable coalgebras the final one.

1 Introduction

An indifferent belief about the bias of a coin can be expressed in various ways. One way is to use elementary sentences like

"When we toss the coin 7 times, the probability of 5 heads is $\frac{1}{8}$."
"When we toss the coin k times, the probability of n heads is $\frac{1}{k+1}$." (1)

Another way is to give a probability measure on the space $[0, 1]$ of biases:

"The bias of the coin is uniformly distributed across the interval $[0, 1]$." (2)

This sentence (2) is more conceptually challenging than (1), but we may deduce (1) from (2) using Lebesgue integration:

$$\int_0^1 \binom{7}{5} r^5 (1 - r)^2 \, dr = \tfrac{1}{8}, \qquad \int_0^1 \binom{k}{n} r^n (1 - r)^{k-n} \, dr = \tfrac{1}{k+1}$$

where $\binom{k}{n} = \frac{k!}{n!(k-n)!}$ is the binomial coefficient. Here we have considered indifferent beliefs as a first illustration, but more generally de Finetti's theorem gives a passage from sentences about beliefs like (1) to sentences like (2).

Theorem 1 (De Finetti, Informal). *If we assign a probability to all finite experiment outcomes involving a coin (such as (1)), and these probabilities are all suitably consistent, then by doing so we have given a probability measure on $[0, 1]$ (such as the uniform distribution, (2)).*

In this paper we give two categorical/coalgebraic formulations of this theorem.

© IFIP International Federation for Information Processing 2020
Published by Springer Nature Switzerland AG 2020
D. Petrişan and J. Rot (Eds.): CMCS 2020, LNCS 12094, pp. 90–111, 2020.
https://doi.org/10.1007/978-3-030-57201-3_6

- In Theorem 8, we state the result as: *the unit interval* [0, 1] *is an inverse limit over a chain of finite spaces in a category of probability channels.* In particular, a cone over this diagram is an assignment of finite probabilities such as (1) that is suitably consistent.
- In Theorem 12, we state the result as: [0, 1] carries a final object in a category of (exchangeable) coalgebras for the functor $2 \times (-)$ in the category of probability channels. A coalgebra can be thought of as a process that coinductively generates probabilities of the form (1). A coalgebra is a little more data than strictly necessary, but this extra data corresponds to the concept of sufficient statistics which is useful in Bayesian inference.

Multisets and Probability Distributions. A multiset is a 'set' in which elements may occur multiple times. We can write such a multiset over a set $\{R, G, B\}$, representing red, green and blue balls, as:

$$3|R\rangle + 5|G\rangle + 2|B\rangle.$$

This multiset can be seen as an urn containing three red, five green and two blue balls.

A probability distribution is a convex combination of elements, where the frequencies (or probabilities) with which elements occur are numbers in the unit interval [0, 1] that add up to one, as in:

$$\tfrac{3}{10}|R\rangle + \tfrac{1}{2}|G\rangle + \tfrac{1}{5}|B\rangle.$$

Obviously, distributions, whether discrete (as above) or continuous, play a central role in probability theory. Multisets also play a crucial role, for instance, an urn with coloured balls is very naturally represented as a multiset. Similarly, multiple data items in learning scenarios are naturally captured by multisets (see *e.g.* [12]). The two concepts are related, since any multiset gives rise to a probability distribution, informally by thinking of it as an urn and then drawing from it (sampling with replacement).

This paper develops another example of how the interplay between multisets and distributions involves interesting structure, in an emerging area of categorical probability theory.

Beliefs as Cones. In brief, following an experiment involving K tosses of a coin, we can set up an urn that approximately simulates the coin according to the frequencies that we have seen. So if we see 5 heads out of 7 tosses, we put 5 black balls into an urn together with 2 white balls, so that the chance of black (=heads) is $\frac{5}{7}$. We can describe a belief about a coin by describing the probabilities of ending up with certain urns (multisets) after certain experiments. Clearly, if we draw and discard a ball from an urn, then this is the same as stopping the experiment one toss sooner. So a consistent belief about our coin determines a cone over the chain

$$\mathcal{M}[1](2) \longleftarrow\!\!\circ\!\!\longrightarrow \mathcal{M}[2](2) \longleftarrow\!\!\circ\!\!\longrightarrow \mathcal{M}[3](2) \longleftarrow\!\!\circ\!\!\longrightarrow \ldots \tag{3}$$

in the category of finite sets and channels (aka probability kernels). Here $\mathcal{M}[K](2)$ is the set of multisets over 2 of size K, *i.e.* urns containing K balls coloured black and white, and the leftward arrows describe randomly drawing and discarding a ball. We use the notation $\multimap$ for channels, a.k.a. Kleisli map, see below for details. Our formulation of de Finetti's theorem says that the limit of this chain is $[0,1]$. So to give a belief, *i.e.* a cone over (3), is to give a probability measure on $[0,1]$. We set up the theory of multisets and cones in Sect. 3, proving the theorem in Sect. 6.

Exchangeable Coalgebras. In many situations, it is helpful to describe a belief about a coin in terms of a stateful generative process, *i.e.* a channel (aka probability kernel) $h\colon X \multimap 2 \times X$, for some space X, which gives coinductively a probability for a sequence of outcomes. This is a coalgebra for the functor $2 \times (-)$. An important example is Pólya's urn, which is an urn-based process, but one where the urn changes over time. Another important example puts the carrier set $X = [0,1]$ as the set of possible biases, and the process merely tosses a coin with the given bias. We show that $[0,1]$ carries the final object among the 'exchangeable' coalgebras. Thus there is a canonical channel $X \multimap [0,1]$, taking the belief described by a generative process such as Pólya's urn to a probability measure on $[0,1]$ such as the uniform distribution. We set up the theory of exchangeable coalgebras in Sect. 4, proving the final coalgebra theorem in Sect. 7 by using the limit reformulation of de Finetti's theorem.

2   Background on Mulitsets and Distributions

Multisets. A *multiset* (or *bag*) is a 'set' in which elements may occur multiple times. We shall write $\mathcal{M}(X)$ for the set of (finite) multisets with elements from a set X. Elements $\varphi \in \mathcal{M}(X)$ may be described as functions $\varphi\colon X \to \mathbb{N}$ with finite support, that is, with $\mathrm{supp}(\varphi) := \{x \in X \mid \varphi(x) \neq 0\}$ is finite. Alternatively, multisets can be described as formal sums $\sum_i n_i |x_i\rangle$, where $n_i \in \mathbb{N}$ describes the multiplicity of $x_i \in X$, that is, the number of times that the element x_i occurs in the multiset. The mapping $X \mapsto \mathcal{M}(X)$ is a monad on the category **Sets** of sets and functions. In fact, $\mathcal{M}(X)$ with pointwise addition and empty multiset $\mathbf{0}$, is the free commutative monoid on X.

Frequently we need to have a grip on the total number of elements occurring in a multiset. Therefore we write, for $K \in \mathbb{N}$,

$$\mathcal{M}[K](X) := \{\varphi \in \mathcal{M}(X) \mid \varphi \text{ has precisely } K \text{ elements}\}$$
$$= \{\varphi \in \mathcal{M}(X) \mid \textstyle\sum_{x \in X} \varphi(x) = K\}.$$

Clearly, $\mathcal{M}[0](X)$ is a singleton, containing only the empty multiset $\mathbf{0}$. Each $\mathcal{M}[K](X)$ is a quotient of the set of sequences X^K via an accumulator map:

$$X^K \xrightarrow{\ acc_X\ } \mathcal{M}[K](X) \quad \text{with} \quad acc_X(x_1,\dots,x_K) := \sum_{x \in X} \left(\textstyle\sum_i \mathbf{1}_{x_i}(x)\right)|x\rangle \quad (4)$$

where $\mathbf{1}_{x_i}$ is the indicator function sending $x \in X$ to 1 if $x = x_i$ and to 0 otherwise. In particular, for $X = 2$, $acc_2(b_1, \ldots, b_K) = (\sum_i b_i)|1\rangle + (K - \sum_i b_i)|0\rangle$. The map acc_X is permutation-invariant, in the sense that:

$$acc_X(x_1, \ldots, x_K) = acc_X(x_{\sigma(1)}, \ldots, x_{\sigma(K)}) \qquad \text{for each permutation } \sigma \in S_K.$$

Such permutation-invariance, or exchangeability, plays an important role in de Finetti's work [5], see the end of Sect. 7.

Discrete Probability. A *discrete probability distribution* on a set X is a finite formal sum $\sum_i r_i |x_i\rangle$ of elements $x_i \in X$ with multiplicity $r_i \in [0, 1]$ such that $\sum_i r_i = 1$. It can also be described as a function $X \to [0, 1]$ with finite support. We write $\mathcal{D}(X)$ for the set of distributions on X. This gives a monad $\mathcal{D}$ on **Sets**; it satisfies $\mathcal{D}(1) \cong 1$ and $\mathcal{D}(2) \cong [0, 1]$, where $1 = \{0\}$ and $2 = \{0, 1\}$.

Maps in the associated Kleisli category $\mathcal{K}\ell(\mathcal{D})$ will be called *channels* and are written with a special arrow $\rightarrow$. Thus $f \colon X \rightarrow Y$ denotes a function $f \colon X \to \mathcal{D}(Y)$. For a distribution $\omega \in \mathcal{D}(X)$ and a channel $g \colon Y \rightarrow Z$ we define a new distribution $f =\!\!\ll \omega \in \mathcal{D}(Y)$ and a new channel $g \circ f \colon X \rightarrow Z$ by Kleisli composition:

$$f =\!\!\ll \omega := \sum_{y \in Y} \left(\sum_x \omega(x) \cdot f(x)(y) \right) |y\rangle \qquad (g \circ f)(x) := g =\!\!\ll f(x).$$

Discrete Probability over Multisets. For each number $K \in \mathbb{N}$ and probability $r \in [0, 1]$ there is the familiar *binomial* distribution $binom[K](r) \in \mathcal{D}(\{0, 1, \ldots, K\})$. It captures probabilities for iterated flips of a known coin, and is given by the convex sum:

$$binom[K](r) := \sum_{0 \le k \le K} \binom{K}{k} \cdot r^k \cdot (1 - r)^{K-k} |k\rangle.$$

The multiplicity probability before $|k\rangle$ in this expression is the chance of getting k heads of out K coin flips, where each flip has fixed bias $r \in [0, 1]$. In this way we obtain a channel $binom[K] \colon [0, 1] \rightarrow \{0, 1, \ldots, K\}$.

More generally, one can define a multinomial channel:

$$\mathcal{D}(X) \xrightarrow{\quad mulnom[K] \quad} \mathcal{M}[K](X).$$

It is defined as:

$$mulnom[K](\omega) := \sum_{\varphi \in \mathcal{M}[K](X)} \frac{K!}{\prod_x \varphi(x)!} \cdot \prod_x \omega(x)^{\varphi(x)} |\varphi\rangle.$$

The binomial channel is a special case, since $[0, 1] \cong \mathcal{D}(2)$ and $\{0, 1, \ldots, K\} \cong \mathcal{M}[K](2)$, via $k \mapsto k|0\rangle + (K - k)|1\rangle$. The binary case will play an important role in the sequel.

3   Drawing from an Urn

Recall our motivation. We have an unknown coin, and we are willing to give probabilities to the outcomes of different finite experiments with it. From this we ultimately want to say something about the coin. The only experiments we consider comprise tossing a coin a fixed number K times and recording the number of heads. The coin has no memory, and so the order of heads and tails doesn't matter. So we can imagine an urn containing black balls and white balls, and in an experiment we start with an empty urn and put a black ball in the urn for each head and a white ball for each tail. The urn forgets the order in which the heads and tails appear. We can then describe our belief about the coin in terms of our beliefs about the probability of certain urns occurring in the different experiments. These probabilities should be consistent in the following informal sense, that connects the different experiments.

- If we run the experiment with $K + 1$ tosses, and then discard a random ball from the urn, it should be the same as running the experiment with just K tosses.

In more detail: because the coin tosses are exchangeable, that is, the coin is memoryless, to run the experiment with $K + 1$ tosses, is to run the experiment with $K + 1$ tosses and then report some chosen permutation of the results. This justifies representing the outcome of the experiment with an unordered urn. It follows that the statistics of the experiment are the same if we choose a permutation of the results at random. To stop the experiment after K tosses, i.e. to discard the ball corresponding to the last toss, is thus to discard a ball uniformly at random.

As already mentioned: an urn filled with certain objects of the kind $x_1, \ldots, x_n$ is very naturally described as a multiset $\sum_i n_i |x_i\rangle$, where n_i is the number of objects x_i in the urn. Thus an urn containing objects from a set X is a multiset in $\mathcal{M}(X)$. For instance, an urn with three black and two white balls can be described as a multiset $3|B\rangle + 2|W\rangle \in \mathcal{M}(2)$, where B stands for black and W for white. We can thus formalise a belief about the K-th experiment, where we toss our coin K times, as a distribution $\omega_K \in \mathcal{D}(\mathcal{M}[K](2))$, on urns with K black and white balls.

For the consistency property, we need to talk about drawing and discarding from an urn. We define a channel that draws a single item from an urn:

$$\mathcal{M}[K+1](X) \xrightarrow{D} \mathcal{D}\Big(X \times \mathcal{M}[K](X)\Big) \tag{5}$$

It is defined as:

$$D(\varphi) := \sum_{x \in \mathrm{supp}(\varphi)} \frac{\varphi(x)}{\sum_y \varphi(y)} |x, \varphi - 1|x\rangle\rangle.$$

Concretely, in the above example:

$$D\big(3|B\rangle + 2|W\rangle\big) = \tfrac{3}{5}|B, 2|B\rangle + 2|W\rangle\rangle + \tfrac{2}{5}|W, 3|B\rangle + 1|W\rangle\rangle.$$

Notice that on the right-hand-side we use brackets $|-\rangle$ inside brackets $|-\rangle$. The outer $|-\rangle$ are for $\mathcal{D}$ and the inner ones for $\mathcal{M}$.

We can now describe the 'draw-and-delete' channel, by post-composing D with the second marginal:

$$\mathcal{M}[K+1](X) \xrightarrow{\quad DD := \mathcal{D}(\pi_2) \circ D \quad} \mathcal{D}\Big(\mathcal{M}[K](X)\Big)$$

Explicitly, $DD(\varphi) = \sum_x \frac{\varphi(x)}{\sum_y \varphi(y)} |\varphi - 1|x\rangle\rangle$.

Now the consistency property requires that the beliefs about the coin comprise a sequence of distributions $\omega_K \in \mathcal{D}(\mathcal{M}[K](2))$ such that $(DD =\!\!\ll \omega_{K+1}) = \omega_K$. Thus they should form a cone over the sequence

$$\mathcal{M}[0](2) \xleftarrow{\ DD\ } \mathcal{M}[1](2) \xleftarrow{\ DD\ } \mathcal{M}[2](2) \xleftarrow{\ DD\ } \ldots \xleftarrow{\ DD\ } \mathcal{M}[K](2) \xleftarrow{\ DD\ } \ldots$$
(6)

This means that for all K,

$$\omega_K(\varphi) = \frac{\varphi(B)+1}{K+1} \cdot \omega_{K+1}(\varphi + |B\rangle) + \frac{\varphi(W)+1}{K+1} \cdot \omega_{K+1}(\varphi + |W\rangle). \qquad (7)$$

Known Bias is a Consistent Belief. If we know for sure that the bias of a coin is $r \in [0,1]$, then we would put $\omega_K = binom[K](r)$. This is a consistent belief, which is generalised in the following result to the multinomial case. This cone forms the basis for this paper.

Proposition 2. *The multinomial channels form a cone for the chain of draw-and-delete channels, as in:*

(8)

Our reformulation of the de Finetti theorem explains the sense in which above cone is a limit when $X = 2$, see Theorem 8.

Proof. For $\omega \in \mathcal{D}(X)$ and $\varphi \in \mathcal{M}[K](X)$ we have:

$$\begin{aligned}
&\big(DD \circ mulnom[K+1]\big)(\omega)(\varphi) \\
&= \sum_\psi mulnom[K+1](\omega)(\psi) \cdot DD[K](\psi)(\varphi) \\
&= \sum_x mulnom[K+1](\omega)(\varphi + 1|x\rangle)) \cdot \frac{\varphi(x)+1}{K+1} \\
&= \sum_x \frac{(K+1)!}{\prod_y (\varphi+1|x\rangle)(y)!} \cdot \prod_y \omega(y)^{(\varphi+1|x\rangle)(y)} \cdot \frac{\varphi(x)+1}{K+1} \\
&= \sum_x \frac{K!}{\prod_y \varphi(y)!} \cdot \prod_y \omega(y)^{\varphi(y)} \cdot \omega(x) \\
&= \frac{K!}{\prod_y \varphi(y)!} \cdot \prod_y \omega(y)^{\varphi(y)} \cdot \big(\sum_x \omega(x)\big) \\
&= mulnom[K](\omega)(\varphi). \qquad\qquad \square
\end{aligned}$$

Unknown Bias (Uniform) is a Consistent Belief. If we do not know the bias of our coin, it may be reasonable to say that all the $(K + 1)$ outcomes in $\mathcal{M}[K](2)$ are equiprobable. This determines the discrete uniform distribution $\upsilon_K \in \mathcal{D}(\mathcal{M}[K](2))$,

$$\upsilon_K := \sum_{k=0}^{K} \tfrac{1}{K+1} |k| B \rangle + (K - k)| W \rangle \rangle.$$

Proposition 3. *The discrete uniform distributions υ_K form a cone for the chain of draw-and-delete channels, as in:*

$$\mathcal{M}[0](X) \xleftarrow{\;DD\;} \cdots \xleftarrow{\;DD\;} \mathcal{M}[K](X) \xleftarrow{\;DD\;} \mathcal{M}[K+1](X) \xleftarrow{\;DD\;} \cdots \tag{9}$$

Proof. For $\varphi \in \mathcal{M}[K](2)$ we have:

$$
\begin{aligned}
\big(DD =\!\!\ll \upsilon_{K+1}\big)(\varphi) &= \sum_{\psi} \upsilon_{K+1}(\psi) \cdot DD[K](\psi)(\varphi) \\
&= \upsilon_{K+1}(\varphi + |W\rangle) \cdot DD[K](\varphi + |W\rangle)(\varphi) + \\
&\qquad \upsilon_{K+1}(\varphi + |B\rangle) \cdot DD[K](\varphi + |B\rangle)(\varphi) \\
&= \tfrac{1}{K+2} \cdot \tfrac{\varphi(W)+1}{K+1} + \tfrac{1}{K+2} \cdot \tfrac{\varphi(B)+1}{K+1} \\
&= \tfrac{\varphi(W)+\varphi(B)+2}{(K+2)(K+1)} \\
&= \tfrac{1}{K+1} \qquad \text{since } \varphi(W) + \varphi(B) = K. \qquad \square
\end{aligned}
$$

4   Coalgebras and Pólya's Urn

In the previous section we have seen two examples of cones over the chain (6): one for a coin with known bias (Proposition 2), and one where the bias of the coin is uniformly distributed (Proposition 3).

In practice, it is helpful to describe our belief about the coin in terms of a data generating process. This is formalised as a channel $h \colon X \to 2 \times X$, *i.e.* as a coalgebra for the functor $2 \times (-)$ on the category of channels. The idea is that our belief about the coin is captured by some parameters, which are the states of the coalgebra $x \in X$. The distribution $h(x)$ describes our belief about the outcome of the next coin toss, but also provides a new belief state according to the outcome. For example, if we have uniform belief about the bias of the coin, we predict the outcome to be heads or tails equally, but afterward the toss our belief is refined: if we saw heads, we believe the coin to be more biased towards heads.

Pólya's urn is an important example of such a coalgebra. We describe it as a multiset over $2 = \{0,1\}$ with a certain extraction and re-fill operation.

The number 0 corresponds to black balls, and 1 to white ones, so that a multiset $\varphi \in \mathcal{M}(2)$ captures an urn with $\varphi(0)$ black and $\varphi(1)$ white ones. When a ball of color $x \in 2$ is extracted, it is put back together with an extra ball of the same color. We choose to describe it as a coalgebra on the set $\mathcal{M}_*(2)$ of non-empty multisets over 2. Explicitly, $pol\colon \mathcal{M}_*(2) \rightarrowtail 2 \times \mathcal{M}_*(2)$ is given by:

$$pol(\varphi) := \frac{\varphi(0)}{\varphi(0) + \varphi(1)} \big| 0, \varphi + 1|0\rangle \big\rangle + \frac{\varphi(1)}{\varphi(0) + \varphi(1)} \big| 1, \varphi + 1|1\rangle \big\rangle.$$

So the urn keeps track of the outcomes of the coin tosses so far, and our belief about the bias of the coin is updated according to successive experiments.

4.1 Iterating Draws from Coalgebras

By iterating a coalgebra $h\colon X \rightarrowtail 2 \times X$, we can simulate the experiment with K coin tosses $h_K\colon X \rightarrowtail \mathcal{M}[K](2)$, by repeatedly applying h. Formally, we build a sequence of channels $h^K\colon X \rightarrowtail \mathcal{M}[K](2) \times X$ by $h^0(x) := (\mathbf{0}, x)$ and:

$$h^{K+1} := X \xrightarrow{h^K} \mathcal{M}[K](2) \times X \xrightarrow{\mathrm{id} \times h} \mathcal{M}[K](2) \times 2 \times X \xrightarrow{\mathrm{add} \times \mathrm{id}} \mathcal{M}[K+1](2) \times X$$

where the rightmost map $add\colon \mathcal{M}[K](2) \times 2 \to \mathcal{M}[K+1](2)$ is $add(\varphi, b) = \varphi + 1|b\rangle$.

(Another way to see this is to regard a coalgebra $h\colon X \rightarrowtail 2 \times X$ in particular as coalgebra $X \rightarrowtail \mathcal{M}(2) \times X$. Now $\mathcal{M}(2) \times -$ is the writer monad for the commutative monoid $\mathcal{M}(2)$, i.e. the monad for $\mathcal{M}(2)$-actions. Iterated Kleisli composition for this monad gives a sequence of maps $h\colon X \rightarrowtail \mathcal{M}(2) \times X$.)

For example, starting with Pólya's urn $pol\colon \mathcal{M}_*(2) \rightarrowtail 2 \times \mathcal{M}_*(2)$, we look at a few iterations:

$$\big(pol\big)^0(b|0\rangle + w|1\rangle) = 1\big|\mathbf{0}, b|0\rangle + w|1\rangle\big\rangle$$
$$\big(pol\big)^1(b|0\rangle + w|1\rangle) = pol(b|0\rangle + w|1\rangle)$$
$$= \tfrac{b}{b+w}\big|1|0\rangle, \varphi + |0\rangle\big\rangle + \tfrac{w}{b+w}\big|1|1\rangle, \varphi + |1\rangle\big\rangle$$
$$\big(pol\big)^2(b|0\rangle + w|1\rangle) = \tfrac{b}{b+w} \cdot \tfrac{b+1}{b+w+1}\big|2|0\rangle, (b+2)|0\rangle + w|1\rangle\big\rangle$$
$$+ \tfrac{b}{b+w} \cdot \tfrac{w}{b+w+1}\big|1|0\rangle + 1|1\rangle, (b+1)|0\rangle + (w+1)|1\rangle\big\rangle$$
$$+ \tfrac{w}{b+w} \cdot \tfrac{b}{b+w+1}\big|1|0\rangle + 1|1\rangle, (b+1)|0\rangle + (w+1)|1\rangle\big\rangle$$
$$+ \tfrac{w}{b+w} \cdot \tfrac{w+1}{b+w+1}\big|2|1\rangle, b|0\rangle + (w+2)|1\rangle\big\rangle$$
$$= \tfrac{b(b+a)}{(b+w)(b+w+1)}\big|2|0\rangle, (b+2)|0\rangle + w|1\rangle\big\rangle$$
$$+ \tfrac{2bw}{(b+w)(b+w+1)}\big|1|0\rangle + 1|1\rangle, (b+1)|0\rangle + (w+1)|1\rangle\big\rangle$$
$$+ \tfrac{w(w+1)}{(b+w)(b+w+1)}\big|2|1\rangle, b|0\rangle + (w+2)|1\rangle\big\rangle.$$

$$(10)$$

It is not hard to see that we get as general formula, for $K \in \mathbb{N}$:

$$\big(pol\big)^K(b|0\rangle + w|1\rangle) = \sum_{0 \le k \le K} \binom{K}{k} \cdot \frac{\prod_{i<k}(b+i) \cdot \prod_{i<K-k}(w+i)}{\prod_{i<K}(b+w+i)}$$
$$\big|k|0\rangle + (K-k)|1\rangle, (b+k)|0\rangle + (w+(K-k))|1\rangle\big\rangle.$$

We are especially interested in the first component of $h^K(x)$, which is in $\mathcal{M}[K](2)$. Thus we define the channels:

$$h_K := \left(X \xrightarrow{\;h^K\;} \mathcal{M}[K](2) \times X \xrightarrow{\;\pi_1\;} \mathcal{M}[K](2) \right)$$

For the Pólya urn pol, the really remarkable fact about $pol_K \colon \mathcal{M}_*(2) \dashrightarrow \mathcal{M}[K](2)$ is that if we start with the urn with one white ball and one black ball, these distributions on $\mathcal{M}[K](2)$ that come from iteratively drawing are uniform distributions:

$$pol_K(1|0\rangle + 1|1\rangle) = \upsilon_K.$$

More generally:

Proposition 4

$$pol_K(b|0\rangle + w|1\rangle)(k|0\rangle + (K-k)|1\rangle)$$
$$= \binom{K}{k} \cdot \frac{(b+k-1)!}{(b-1)!} \cdot \frac{(w+K-k-1)!}{(w-1)!} \cdot \frac{(b+w-1)!}{(b+w+K-1)!}. \tag{11}$$

Proof. We start from the K-th iteration formula for pol, given just before Proposition 4, and derive:

$$\frac{\prod_{i<k}(b+i) \cdot \prod_{i<K-k}(w+i)}{\prod_{i<K}(b+w+i)}$$
$$= \frac{(b+k-1)!}{(b-1)!} \cdot \frac{(w+K-k-1)!}{(w-1)!} \cdot \frac{(b+w-1)!}{(b+w+K-1)!}. \qquad \square$$

From this characterization, we can deduce that the channels $pol_K \colon \mathcal{M}_*(2) \dashrightarrow \mathcal{M}[K](2)$ form a cone over the draw-and-delete chain (6), *i.e.* that $DD \circ pol_{K+1} = pol_K$, by a routine calculation.

In what follows we provide two other ways to see that pol_K forms a cone. First we deduce it using the fact that the coalgebra pol is exchangeable (Lemma 6). Second we deduce it by noticing that $pol_K(b|0\rangle + w|1\rangle)(k|0\rangle + (K-k)|1\rangle)$ is distributed as $betabin[K](b,w)$, the beta binomial distribution, and hence factors through the binomial cone (Proposition 2) via the beta channel (Lemma 7).

4.2   Exchangeable Coalgebras and Cones

We would like to focus on those coalgebras $h \colon X \dashrightarrow 2 \times X$ that describe consistent beliefs, *i.e.* for which $h_K \colon X \dashrightarrow \mathcal{M}[K](2)$ is a cone over the chain (6):

$$DD \circ h_{K+1} = h_K. \tag{12}$$

For example, Pólya's urn forms a cone. But the deterministic alternating coalgebra

$$alt : 2 \dashrightarrow 2 \times 2 \qquad alt(b) := 1|b, \neg b\rangle \tag{13}$$

does not form a cone, since for instance $alt_1(0) = 1|0\rangle$, while $DD \circ alt_2(0) = \frac{1}{2}|0\rangle + \frac{1}{2}|1\rangle$.

The following condition is inspired by de Finetti's notion of exchangeability, and gives a simple coinductive criterion for a coalgebra to yield a cone (12). It is also related to ideas from probabilistic programming (*e.g.* [1]).

Definition 5. *A coalgebra* $h\colon X \nrightarrow 2 \times X$ *will be called* exchangeable *if its outputs can be reordered without changing the statistics, as expressed by commutation of the following diagram.*

$$
\begin{array}{ccc}
\overset{h}{\nrightarrow} 2 \times X & \xrightarrow{\;id\otimes h\;} & 2 \times 2 \times X \\
& & \\
X & \cong \Big\downarrow swap \times id_X & \quad (14) \\
& & \\
\underset{h}{\nrightarrow} 2 \times X & \xrightarrow[\;id\otimes h\;]{} & 2 \times 2 \times X
\end{array}
$$

Returning to Pólya's urn coalgebra $pol\colon \mathcal{M}_*(2) \nrightarrow 2 \times \mathcal{M}_*(2)$, we see that it is exchangeable, for, by a similar argument to (10),

$$
\begin{aligned}
(id \times pol)(pol(b|0\rangle + w|1\rangle))) = {} = {} & \tfrac{b}{b+w} \cdot \tfrac{b+1}{b+w+1} \big| (0,0), (b+2)|0\rangle + w|1\rangle \big\rangle \\
& + \tfrac{b}{b+w} \cdot \tfrac{w}{b+w+1} \big| (0,1), (b+1)|0\rangle + (w+1)|1\rangle \big\rangle \\
& + \tfrac{w}{b+w} \cdot \tfrac{b}{b+w+1} \big| (1,0), (b+1)|0\rangle + (w+1)|1\rangle \big\rangle \\
& + \tfrac{w}{b+w} \cdot \tfrac{w+1}{b+w+1} \big| (1,1), b|0\rangle + (w+2)|1\rangle \big\rangle
\end{aligned}
$$

Reordering the results does not change the statistics, because $b \cdot w = w \cdot b$. On the other hand, the deterministic alternating coalgebra $alt\colon 2 \nrightarrow 2 \times 2$ from (13) is not exchangeable, since:

$$
\begin{aligned}
(id \times alt)(alt(0)) &= |((0,1),0)\rangle \\
(swap \times 2)((id \times alt)(alt(0))) &= |((1,0),0)\rangle.
\end{aligned}
$$

Before we state and prove our theorem about exchangeable coalgebras, we also consider a different sequence, which keeps track of the order of draws. We define a sequence $h^{\sharp K}\colon X \nrightarrow 2^K \times X$, with $h^{\sharp 0}(x) = ((), x)$ and $h^{\sharp(K+1)}(x) = (2^K \otimes h) =\!\!\ll h^{\sharp K}(x)$. Then we define channels $h_K^\sharp \colon X \nrightarrow 2^K$ by

$$
h_K^\sharp(x) := \mathcal{D}(\pi_1)(h^{\sharp K}(x)) \tag{15}
$$

These constructions are related similar to the earlier constructions h^K and h_K by

$$
h^K = \left(X \xrightarrow{\;h^{\sharp K}\;} 2^K \times X \xrightarrow{\;acc \times X\;} \mathcal{M}[K](2) \times X \right)
$$

$$
h_K = \left(X \xrightarrow{\;h_K^\sharp\;} 2^K \xrightarrow{\;\;acc\;\;} \mathcal{M}[K](2) \right),
$$

where $acc\colon 2^K \to \mathcal{M}[K](2)$ is the accumulator map from (4). The difference is that the $h^\sharp$ variants keep track of the order of draws, which will be useful in our arguments about exchangeability, since h is exchangeable (Definition 5) if and only if $h^{\#2} = \mathcal{D}(\text{swap} \times X) \circ h^{\#2} : X \dashrightarrow 2 \times 2 \times X$.

Lemma 6. *Let* $h\colon X \dashrightarrow 2 \times X$ *be an exchangeable coalgebra. Then*

1. *the map* h_K *can be expressed as:*

$$h_K(x) = \sum_{k=0}^{K} \binom{K}{k} \cdot h_K^{\#}(x)\big(\ \underbrace{1,\ldots,1}_{k\ times},\ \underbrace{0,\ldots,0}_{K-k\ times}\ \big) \,|\, k|1\rangle + (K-k)|0\rangle\rangle. \quad (16)$$

2. *the collection of maps* (h_K) *forms a cone for the chain of draw-and-delete maps* DD (6).

Proof. 1. Among $b_1, \ldots, b_K \in 2$ there are $\binom{K}{k}$-many possibilities of having k-many ones. By exchangeability of h there is a normal form with all these ones upfront

$$h_K^{\#}(x)\big(b_1, \ldots, b_K\big) = h_K^{\#}(x)\big(\ \underbrace{1,\ldots,1}_{k\ times},\ \underbrace{0,\ldots,0}_{K-k\ times}\ \big)$$

2. Using (7) we get, for any $\varphi = (k|1\rangle + (K-k)|0\rangle) \in \mathcal{M}[K](2)$,

$$
\begin{aligned}
&\big(DD \circ h_{K+1}\big)(x)(\varphi) \\
&= \tfrac{k+1}{K+1} \cdot h_{K+1}(x)(\varphi + |1\rangle) + \tfrac{K+1-k}{K+1} \cdot h_{K+1}(\varphi + |0\rangle) \\
&\overset{(16)}{=} \tfrac{k+1}{K+1} \cdot \binom{K+1}{k+1} \cdot h_{K+1}^{\#}(x)\big(\ \underbrace{1,\ldots,1}_{k+1\ times},0,\ldots,0\big) + \\
&\qquad \tfrac{K+1-k}{K+1} \cdot \binom{K+1}{k} \cdot h_{K+1}^{\#}(x)\big(\underbrace{1,\ldots,1}_{k\ times},0,\ldots,0\big) \\
&= \binom{K}{k} \cdot \Big(h_{K+1}^{\#}(x)\big(\underbrace{1,\ldots,1}_{k\ times},0,\ldots,0,1\big) + h_{K+1}^{\#}(x)\big(\underbrace{1,\ldots,1}_{k\ times},0,\ldots,0\big)\Big) \\
&= \binom{K}{k} \cdot \textstyle\sum_y h^K(x)\big(\underbrace{1,\ldots,1}_{k\ times},0,\ldots,0,y\big) \cdot \Big(h_1^{\#}(y)(1) + h_1^{\#}(y)(0)\Big) \\
&= \binom{K}{k} \cdot \textstyle\sum_y h^K(x)\big(\underbrace{1,\ldots,1}_{k\ times},0,\ldots,0,y\big) \cdot 1 \\
&= \binom{K}{k} \cdot h_K^{\#}(x)\big(\underbrace{1,\ldots,1}_{k\ times},0,\ldots,0\big) \\
&\overset{(16)}{=} h_K(x)(\varphi).
\end{aligned}
$$

$\square$

Since the Pólya's urn coalgebra *pol* is exchangeable, the lemma provides another way to deduce that the channels $pol_K\colon \mathcal{M}_*(2) \dashrightarrow \mathcal{M}[K](2)$ form a cone (12).

5   Background on Continuous Probability

De Finetti's theorem is about the passage from discrete probability to continuous probability. Consider the uniform distribution μ on $[0, 1]$. The probability of any particular point is 0. So rather than give a probability to individual points, we give probabilities to sets; for example, the probability of a point in the half interval $[0, \frac{1}{2}]$ is $\frac{1}{2}$.

In general, a σ-algebra on a set X is a collection of sets Σ that is closed under countable unions and complements. A measurable space (X, Σ) is a set together with a σ-algebra; the sets in Σ are called measurable. In this paper we mainly consider two kinds of σ-algebra, and so they will often be left implicit:

- on a countable set (*e.g.* 2, $\mathcal{M}(2)$, $\mathcal{M}[K](2)$) we consider the σ-algebra where every set is measurable.
- on the unit interval $[0, 1]$, the Borel σ-algebra, which is the least σ-algebra containing the intervals.

On a measurable space (X, Σ), a *probability measure* is a function $\omega\colon \Sigma \to [0, 1]$ with $\omega(X) = 1$ and with $\omega(\bigcup_i U_i) = \sum_i \omega(U_i)$ when the measurable subsets $U_i \in \Sigma$ are pairwise disjoint. On a finite set X the probability measures coincide with the discrete distributions.

The morphisms of measurable spaces are the functions $f\colon X \to Y$ for which the inverse image of a measurable set is measurable. Crucially, if $f\colon X \to [0, 1]$ is measurable and $\omega\colon \Sigma \to [0, 1]$ is a probability measure, then we can find the Lebesgue integral $\int f(x)\,\omega(\mathrm{d}x)$, which is the expected value of f. When X is finite, the Lebesgue integral is a sum: $\int f(x)\,\omega(\mathrm{d}x) = \sum_X f(x)\omega(\{x\})$. When $X = [0, 1]$ and ω is the uniform distribution, then the Lebesgue integral is the familiar one.

We write $\mathcal{G}(X, \Sigma)$ for the set of all probability measures on (X, Σ), where we often omit the σ-algebra Σ when it is clear from the context.: $\mathcal{D}(X) \cong \mathcal{G}(X)$, when all subsets are seen as measurable. The set $\mathcal{G}(X, \Sigma)$ itself has a σ-algebra, which is the least σ-algebra making the sets $\{\omega \mid \omega(U) < r\}$ measurable for all fixed U and r.

Like $\mathcal{D}$, the construction $\mathcal{G}$ is a monad, commonly called the Giry monad, but $\mathcal{G}$ is a monad on the category of measurable spaces. Maps in the associated Kleisli category $\mathcal{K}\ell(\mathcal{G})$ will also be called *channels*. Thus $f\colon X \dashrightarrow Y$ between measurable spaces can be described as a function $f\colon X \times \Sigma_Y \to [0, 1]$ that is a measure in Σ_Y and measurable in X. For a measure $\omega \in \mathcal{G}(X)$ and a channel $g\colon Y \dashrightarrow Z$ we define a new distribution $f =\!\!\ll \omega \in \mathcal{G}(Y)$ and a new channel $g \circ f\colon X \dashrightarrow Z$ by Kleisli composition:

$$(f =\!\!\ll \omega)(U) := \int_X f(x)(U)\,\omega(\mathrm{d}x) \qquad (g \circ f)(x) := g =\!\!\ll f(x).$$

Notice that a discrete channel is in particular a continuous channel, according to our terminology, and so we have an inclusion of Kleisli categories categories $\mathcal{K}\ell(\mathcal{D}) \hookrightarrow \mathcal{K}\ell(\mathcal{G})$.

Illustrations from the Beta-Bernoulli Distributions. At the end of Sect. 4.1, we briefly mentioned the beta-bernoulli distribution as it arises from the coalgebra for Pólya's urn. In general, this beta-binomial channel is defined for arbitrary $\alpha, \beta \in \mathbb{R}_{>0}$ as:

$$betabin[K](\alpha, \beta) := \sum_{0 \leq k \leq K} \binom{K}{k} \cdot \frac{B(\alpha + k, \beta + K - k)}{B(\alpha, \beta)} |k|0\rangle + (K - k)|1\rangle\rangle, \quad (17)$$

This description involves the function $B \colon \mathbb{R}_{>0} \times \mathbb{R}_{>0} \to \mathbb{R}$ defined as:

$$B(\alpha, \beta) := \int_0^1 x^{\alpha-1} \cdot (1 - x)^{\beta-1} \, dx \quad (18)$$

This B is a well-known mathematical function satisfying for instance:

$$B(\alpha + 1, \beta) = \frac{\alpha}{\alpha + \beta} \cdot B(\alpha, \beta) \qquad B(n + 1, m + 1) = \frac{n! \cdot m!}{(n + m + 1)!}, \quad (19)$$

for $n, m \in \mathbb{N}$. The latter equation shows that (11) is an instance of (17).

The function B is also used to define the Beta distribution on the unit interval $[0, 1]$. We describe it here as a channel in the Kleisli category of the Giry monad:

$$\mathbb{R}_{>0} \times \mathbb{R}_{>0} \xrightarrow{\quad beta \quad} \mathcal{G}([0, 1]).$$

It is defined on $\alpha, \beta \in \mathbb{R}_{>0}$ and measurable $M \subseteq [0, 1]$ as:

$$beta(\alpha, \beta)(M) := \int_M \frac{x^{\alpha-1} \cdot (1 - x)^{\beta-1}}{B(\alpha, \beta)} \, dx. \quad (20)$$

We prove some basic properties of the beta-binomial channel. These properties are well-known, but are formulated here in slightly non-standard form, using channels.

Lemma 7. *1. A betabinomial channel can itself be decomposed as:*

$$
\begin{array}{ccc}
 & \mathcal{M}[K](2) & \\
betabin[K] \nearrow & & \uparrow binom[K] \\
\mathbb{R}_{>0} \times \mathbb{R}_{>0} & \xrightarrow{\quad beta \quad} & [0, 1]
\end{array}
$$

2. The beta-binomial channels

$$\mathbb{R}_{>0} \times \mathbb{R}_{>0} \xrightarrow{\quad betabin[K] \quad} \mathcal{M}[K](2)$$

for $K \in \mathbb{N}$, form a cone for the infinite chain of draw-and-delete channels (6).

Proof. 1. Composition in the Kleisli category $\mathcal{K}\ell(\mathcal{G})$ of the Giry monad $\mathcal{G}$ yields:

$$
\begin{aligned}
&\big(binom[K] \circ beta\big)(\alpha, \beta)\big(k|0\rangle + (K - k)|1\rangle\big) \\
&= \int_0^1 \binom{K}{k} \cdot x^k \cdot (1 - x)^{K-k} \cdot \frac{x^{\alpha-1} \cdot (1 - x)^{\beta-1}}{B(\alpha, \beta)} \, \mathrm{d}x \\
&= \binom{K}{k} \cdot \frac{\int_0^1 x^{\alpha+k-1} \cdot (1 - x)^{\beta+(K-k)-1} \, \mathrm{d}x}{B(\alpha, \beta)} \\
&\overset{(18)}{=} \binom{K}{k} \cdot \frac{B(\alpha + k, \beta + K - k)}{B(\alpha, \beta)} \\
&\overset{(17)}{=} betabin[K](\alpha, \beta)\big(k|0\rangle + (K - k)|1\rangle\big).
\end{aligned}
$$

2. The easiest way to see this is by observing that the channels factorise via the binomial channels $binom[K] \colon [0, 1] \dashrightarrow \mathcal{M}[K](2)$. In Proposition 2 we have seen that these binomial channels commute with draw-and-delete channels DD. $\qquad\square$

6   De Finetti as a Limit Theorem

We first describe our 'limit' reformulation of the de Finetti theorem and later on argue why this is a reformulation of the original result.

Theorem 8. *For $X = 2$ the cone (8) is a limit in the Kleisli category $\mathcal{K}\ell(\mathcal{G})$ of the Giry monad; this limit is the unit interval $[0, 1] \cong \mathcal{D}(2)$.*

This means that if we have channels $c_K \colon Y \dashrightarrow \mathcal{M}[K](2) \cong \{0, 1, \ldots, K\}$ with $DD \circ c_{K+1} = c_K$, then there is a unique mediating channel $c \colon Y \dashrightarrow [0, 1]$ with $binom[K] \circ c = c_K$, for each $K \in \mathbb{N}$.

In the previous section we have seen an example with $Y = \mathbb{R}_{>0} \times \mathbb{R}_{>0}$ and $c_K = betabin[K] \colon Y \dashrightarrow \mathcal{M}[K](2)$ and mediating channel $c = beta \colon Y \dashrightarrow [0, 1]$.

Below we give the proof, in the current setting. We first illustrate the central role of the draw-and-delete maps $DD \colon \mathcal{M}[K+1](X) \dashrightarrow \mathcal{M}[K](X)$ and how they give rise to hypergeometric and binomial distributions.

Via the basic isomorphism $\mathcal{M}[K](2) \cong \{0, 1, \ldots, K\}$ the draw-and-delete map (7) becomes a map $DD \colon \{0, \ldots, K + 1\} \dashrightarrow \{0, \ldots, K\}$, given by:

$$
DD(\ell) = \tfrac{\ell}{K+1}|\ell - 1\rangle + \tfrac{K+1-\ell}{K+1}|\ell\rangle. \tag{21}
$$

Notice that the border cases $\ell = 0$ and $\ell = K + 1$ are handled implicitly, since the corresponding probabilities are then zero.

We can iterate DD, say N times, giving $DD^N \colon \{0, \ldots, K+N\} \dashrightarrow \{0, \ldots, K\}$. If we draw-and-delete N times from an urn with $K + N$ balls, the result is the same as sampling without replacement K times from the same urn. Sampling without replacement is modelled by the *hypergeometric distribution*. So we can calculate DD^N explicitly, using the hypergeometric distribution formula:

$$
DD^N(\ell) = \sum_{0 \le k \le K} \frac{\binom{\ell}{k} \cdot \binom{K+N-\ell}{K-k}}{\binom{K+N}{K}} |k\rangle.
$$

Implicitly the formal sum is over $k \leq \ell$, since otherwise the corresponding probabilities are zero.

As is well known, for large N, the hypergeometric distribution can be approximated by the binomial distribution. To be precise, for fixed K, if $(\ell/K+N) \to p$ as $N \to \infty$, then the hypergeometric distribution $DD^N(\ell)$ approaches the binomial distribution $binom[K](p)$ as N goes to infinity. This works as follows. Using the notation $(n)_m = \prod_{i<m}(n-i) = n(n-1)(n-2)\cdots(n-m+1)$ we can write $\binom{n}{m} = (n)_m/n!$ and thus:

$$
\begin{aligned}
\frac{\binom{\ell}{k} \cdot \binom{K+N-\ell}{K-k}}{\binom{K+N}{K}} &= \frac{(\ell)_k}{k!} \cdot \frac{(K+N-\ell)_{K-k}}{(K-k)!} \cdot \frac{K!}{(K+N)_K} \\
&= \binom{K}{k} \cdot \frac{(\ell)_k \cdot (K+N-\ell)_{K-k}}{(K+N)_K} \\
&= \binom{K}{k} \cdot \frac{\prod_{i<k}(\ell-i) \cdot \prod_{i<K-k}(K+N-\ell-i)}{\prod_{i<K}(K+N-i)} \\
&= \binom{K}{k} \cdot \frac{\prod_{i<k}(\ell/K+N - i/K+N) \cdot \prod_{i<K-k}(1 - \ell/K+N - i/K+N)}{\prod_{i<K}(1 - i/K+N)} \\
&\to \binom{K}{k} \cdot p^k \cdot (1-p)^{K-k} \qquad \text{as } N \to \infty \\
&= binom[K](p)(k).
\end{aligned}
$$

The proof of Theorem 8 makes use of a classical result of Hausdorff [10], giving a criterion for the existence of probability measures. It is repeated here without proof (but see *e.g.* [6] for details).

Recall that the *moments* of a distribution $\mu \in \mathcal{G}([0,1])$ form a sequence of numbers $\mathfrak{m}_1, \ldots, \mathfrak{m}_K, \ldots \in [0,1]$ given by $\mathfrak{m}_K = \int_0^1 r^K \mu(dr)$: the expected values of r^K. In general, moments correspond to statistical concepts such as mean, variance, skew and so on. But in this specific case, the K-th moment of μ is the probability $(binom[K] =\!\!\ll \mu)(K)$ that K balls drawn will *all* be black.

Recall that a sequence $a = (a_n)$ is *completely monotone* if:

$$
(-1)^k \cdot \left(\Delta^k a\right)_n \geq 0 \qquad \text{for all } n, k.
$$

This formulation involves the difference operator Δ on sequences, given by $(\Delta a)_n = a_{n+1} - a_n$. This operator is iterated, in the standard way as $\Delta^0 = \text{id}$ and $\Delta^{m+1} = \Delta \circ \Delta^m$.

Theorem 9 (Hausdorff). *A sequence $a = (a_n)_{n\in\mathbb{N}}$ of non-negative real numbers a_n is* completely monotone *if and only if it arises as sequence of moments: there is a unique measure $\mu \in \mathcal{G}([0,1])$ with, for each n,*

$$
a_n = \int_0^1 x^n \, \mu(dx). \qquad \square
$$

Proof (of Theorem 8). As a first step, we shall reason pointwise. Let, for each $K \in \mathbb{N}$, a distribution $\omega_K \in \mathcal{D}(\{0, 1, \ldots, K\})$ be given, with $DD =\!\!\ll \omega_{K+1} = \omega_K$.

We need to produce a unique (continuous) distribution $\omega \in \mathcal{G}([0,1])$ such that $\omega_K = binom[K] =\!\!\ll \omega$, for each K.

The problem of finding ω is essentially Hausdorff's moment problem, see Theorem 9. The connection between Hausdorff's moments problem and de Finetti's theorem is well known [6, §VII.4], but the connection is particularly apparent in our channel-based formulation.

We exploit Hausdorff's criterion via two claims, which we elaborate in some detail. The first one is standard.

Claim 1. If $\mu \in \mathcal{G}([0,1])$ has moments $\mathfrak{m}_K$, then:

$$\int_0^1 x^n \cdot (1-x)^k \, \mu(\mathrm{d}x) = (-1)^k \cdot \left(\Delta^k \mathfrak{m}\right)_n.$$

This equation can be proven by induction on k. It holds by definition of $\mathfrak{m}$ for $k = 0$. Next:

$$\int_0^1 x^n \cdot (1-x)^{k+1} \, \mu(\mathrm{d}x)$$
$$= \int_0^1 x^n \cdot (1-x) \cdot (1-x)^k \, \mu(\mathrm{d}x)$$
$$= \int_0^1 x^n \cdot (1-x)^k \, \mu(\mathrm{d}x) - \int_0^1 x^{n+1} \cdot (1-x)^k \, \mu(\mathrm{d}x)$$
$$\overset{(\mathrm{IH})}{=} (-1)^k \cdot \left(\Delta^k \mathfrak{m}\right)_n - (-1)^k \cdot \left(\Delta^k \mathfrak{m}\right)_{n+1}$$
$$= (-1)^{k+1} \cdot \left(\left(\Delta^k \mathfrak{m}\right)_{n+1} - \left(\Delta^k \mathfrak{m}\right)_n\right)$$
$$= (-1)^{k+1} \cdot \left(\Delta^{k+1} \mathfrak{m}\right)_n.$$

The next claim involves a crucial observation about our setting: the whole sequence of distributions $\omega_K \in \mathcal{D}(\{0, 1, \ldots, K\})$ is entirely determined by the probabilities $\omega_K(K)$ of drawing K balls which are all black.

Claim 2. The numbers $\mathfrak{b}_K := \omega_K(K) \in [0,1]$, for $K \in \mathbb{N}$, determine all the distributions ω_K via the relationship:

$$\omega_{k+n}(n) = \binom{k+n}{n} \cdot (-1)^k \cdot \left(\Delta^k \mathfrak{b}\right)_n.$$

In particular, the sequence $\mathfrak{b}_K$ is completely monotone.

The proof works again by induction on k. The case $k = 0$ holds by definition. For the induction step we use the cone property $DD =\!\!\ll \omega_{k+1} = \omega_k$. Via (21) it gives:

$$\omega_k(i) = \frac{i+1}{k+1} \cdot \omega_{k+1}(i+1) + \frac{k+1-i}{k+1} \cdot \omega_{k+1}(i),$$

so that:

$$\omega_{k+1}(i) = \frac{k+1}{k+1-i} \cdot \omega_k(i) - \frac{i+1}{k+1-i} \cdot \omega_{k+1}(i+1).$$

We use this equation in the first step below.

$$\omega_{(k+1)+n}(n)$$
$$\overset{(\text{IH})}{=} \frac{k+1+n}{k+1} \cdot \omega_{k+n}(n) - \frac{n+1}{k+1} \cdot \omega_{k+1+n}(n+1)$$
$$= \frac{k+1+n}{k+1} \cdot \binom{k+n}{n} \cdot (-1)^k \cdot \left(\Delta^k\mathfrak{b}\right)_n - \frac{n+1}{k+1} \cdot \binom{k+1+n}{n+1} \cdot (-1)^k \cdot \left(\Delta^k\mathfrak{b}\right)_{n+1}$$
$$= \binom{k+1+n}{n} \cdot (-1)^{k+1} \cdot \left(\left(\Delta^k\mathfrak{b}\right)_{n+1} - \left(\Delta^k\mathfrak{b}\right)_n\right)$$
$$= \binom{k+1+n}{n} \cdot (-1)^{k+1} \cdot \left(\Delta^{k+1}\mathfrak{b}\right)_n.$$

This second claim implies that there is a unique distribution $\omega \in \mathcal{G}([0,1])$ whose K-th moments equals $\mathfrak{b}_K = \omega_K(K)$. But then:

$$\omega_K(k) = \omega_{(K-k)+k}(k) = \binom{K}{k} \cdot (-1)^{K-k} \cdot \left(\Delta^{K-k}\mathfrak{b}\right)_k \qquad \text{by Claim 2}$$
$$= \binom{K}{k} \cdot \int_0^1 x^k \cdot (1-x)^{K-k} \, d\omega(x) \text{ by Claim 1}$$
$$= \left(binom[K] =\!\!\ll \omega\right)(k).$$

As an aside, it is perhaps instructive to look at how the distribution $\omega \in \mathcal{G}([0,1])$ is built in the proof of Hausdorff's criterion. One way is as a limiting measure of the sequence of finite discrete distributions on $[0,1]$,

$$\overline{\omega}_K := \sum_{k=0}^{K} \omega_K(k) \big| \tfrac{k}{K} \big\rangle. \tag{22}$$

This is illustrated in the following table, for the examples of the binomial cone and the uniform cone. The graphs show $\overline{\omega}_K$ (plotted as densities with respect to the counting measure).

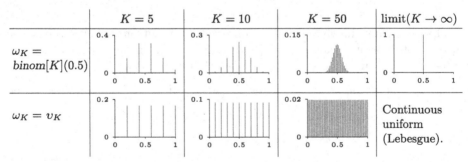

	$K=5$	$K=10$	$K=50$	limit($K \to \infty$)
$\omega_K =$ $binom[K](0.5)$				
$\omega_K = \upsilon_K$				Continuous uniform (Lebesgue).

We conclude by explaining why this construction gives a measurable mediating map. Suppose we have a measurable space (Y, Σ) and a sequence of channels $c_K: Y \rightsquigarrow \{0,\ldots,K\}$ forming a cone over (8), *i.e.* $c_K = DD \circ c_{K+1}$ for all $K \in \mathbb{N}$. We define a mediating channel $c: Y \rightsquigarrow [0,1]$ by putting the distribution $c(y)$ as the unique one with moments $c_K(y)(K)$, as above. It remains to show that c, regarded as a function $Y \to \mathcal{G}([0,1])$, is a measurable function. It is immediate that the sequence-of-moments function $\mathfrak{m}: Y \to [0,1]^\infty$ given by:

$$\mathfrak{m}(y)_K := \int_0^1 r^K \, c(x)(dr) = c_K(y)(K).$$

is measurable, because the c_K's are assumed to be channels: let $V \subseteq [0,1]$ be measurable; for associated basic measurable subsets of $[0,1]^\infty$ one has:

$$\mathfrak{m}^{-1}\big([0,1]^K \times V \times [0,1]^\infty\big) = \{y \in Y \mid c_K(y)(K) \in V\} = c_K^{-1}\big([0,1]^K \times V\big).$$

Further, $c\colon Y \to \mathcal{G}([0,1])$ factors through $\mathfrak{m}$; indeed $\mathcal{G}([0,1])$ can be thought of as a subset of $[0,1]^\infty$, when each distribution is regarded as its sequence of moments. We conclude by the less well-known fact that $\mathcal{G}([0,1])$ is a sub*space* of $[0,1]^\infty$. $\qquad\qquad\square$

Theorem 8 is a reformulation of a classical result of de Finetti [5]. It introduced the concept of *exchangeability* of random variables which has disappeared completely from our reformulation, but is implicit in our use of multisets. This requires some explanation.

A (discrete) random variable is a pair $\rho \in \mathcal{D}(A), U\colon A \to \mathbb{R}$ of a distribution and an observable. Usually the distribution ρ is implicit. Instead, the notation $P(U)$ is used for the image distribution $\mathcal{D}(U)(\rho)$ on $\mathbb{R}$, also sometimes written as $P(U = r)$, for $r \in \mathbb{R}$.

Let $(U_1, \rho_1), \ldots, (U_n, \rho_n)$ be a finite sequence of random variables. It induces a joint distribution on $\mathbb{R}^n$,

$$P(U_1, \ldots, U_n) := \mathcal{D}\big(U_1 \times \cdots \times U_n\big)\big(\rho_1 \otimes \cdots \otimes \rho_n\big),$$

where $U_1 \times \cdots \times U_n\colon A_1 \times \cdots \times A_n \to \mathbb{R}^n$ and $\rho_1 \otimes \cdots \otimes \rho_n \in \mathcal{D}(A_1 \times \cdots \times A_n)$.

Such a sequence (U_i, ρ_i) is called *exchangeable* if for each permutation π:

$$P(U_1, \ldots, U_n) = P(U_{\pi(1)}, \ldots, U_{\pi(n)}).$$

An infinite sequence is called exchangeable if each finite subsequence is exchangeable.

The original form of de Finetti's theorem involves binary (0–1) random variables with $U_i\colon A_i \to 2 = \{0,1\} \hookrightarrow \mathbb{R}$. The distribution $P(U_i)$ is then on a 'Bernouilli' distribution on 2. When random variables are exchangeable their order doesn't matter but only how many of them are zero or one. This is captured by their sum $\sum_i U_i$. But this sum is a *multiset* over 2. Indeed, in a multiset, the order of elements does not matter, only their multiplicities (numbers of occurrences). This is a crucial observation underlying our reformulation, that already applied to the accumulator map *acc* in (4). But let's first look at the usual formulation of de Finetti's (binary) representation theorem, see [5] and the textbooks [6,16].

Theorem 10. *Let $U_1, U_2, \ldots$ be an infinite sequence of exchangeable binary random variables. Then there is a probability measure ω on $[0,1]$ such that for all n,*

$$P(U_1, \ldots, U_n) = \int_0^1 x^{\sum_i U_i}(1 - x)^{(n - \sum_i U_i)}\,\omega(\,\mathrm{d}x).$$

It is clear that there is a considerable gap between this original formulation and the inverse limit formulation that has been developed in this paper. There is room for narrowing this gap.

The essence that is used in de Finetti's theorem about n exchangeable binary random variables are their outcomes via the accumulator map acc from (4), giving a map $\mathcal{D}(acc)\colon \mathcal{D}(2^n) \to \mathcal{D}(\mathcal{M}[n](2))$.

In our approach we have reformulated the requirements about random variables and about exchangeability and worked directly with distributions on multisets, see Proposition 2 and Theorem 8. De Finetti's (binary) representation Theorem 10 becomes a limit statement in a Kleisli category. We think that this approach brings forward the mathematical essentials.

7   A Final Exchangeable Coalgebra

We return to the setting of coalgebras, begun in Sect. 4. Any coalgebra $h\colon X \rightarrow 2 \times X$ for the functor $2 \times (-)$ on $\mathcal{K}\ell(\mathcal{D})$ or on $\mathcal{K}\ell(\mathcal{G})$ is a probabilistic process that produces a random infinite sequence of boolean values. Following [14], we can understand this induced sequence as arising from the fact that the pair of head and tail function

$$2^\infty \xrightarrow{\ \cong\ } 2 \times 2^\infty$$

is a final coalgebra in the category $\mathcal{K}\ell(\mathcal{G})$, and so there is a unique channel $h^\sharp\colon X \rightarrow 2^\infty$ assigning to each element of X the random sequence that it generates. Here 2^∞ is the measurable space of infinite binary streams, regarded with the product σ-algebra.

It is helpful to spell this construction out a little. The sequence of first projections

$$2^0 \longleftarrow 2^1 \longleftarrow 2^2 \longleftarrow 2^3 \longleftarrow \cdots \tag{23}$$

has as a limit the space 2^∞, with the limiting cone being the obvious projections. And this limit is also a limit in $\mathcal{K}\ell(\mathcal{G})$. (This is a categorical formulation of Kolmogorov's extension theorem, see [4, Thm. 2.5].)

Now, as usual in coalgebra theory, we find $h^\sharp\colon X \rightarrow 2^\infty$ for any coalgebra $h\colon X \rightarrow 2 \times X$ by noticing that the channels $h^\sharp_K\colon X \rightarrow 2^K$ (15) define a cone over the chain (23). The limiting channel $h^\sharp\colon X \rightarrow 2^\infty$ for the cone is the unique coalgebra homomorphism to the final coalgebra.

Example 11. We define the *bernoulli coalgebra* to be the channel $bern\colon [0,1] \rightarrow 2 \times [0,1]$ given by $bern(r) := r \cdot |1,r\rangle + (1-r) \cdot |0,r\rangle$. In words, $bern(r)$ makes a coin toss with known bias r and returns the outcome together with the same bias r unchanged.

The final channel $iid := bern^\sharp\colon [0,1] \rightarrow 2^\infty$ takes a probability r to a random infinite sequence of independent and identically distributed binary values, all distributed as $r|1\rangle + (1-r)|0\rangle$. More explicitly, for basic measurable subsets, at each stage n,

$$iid(r)\big(2^n \times \{1\} \times 2^\infty\big) = r \quad \text{and} \quad iid(r)\big(2^n \times \{0\} \times 2^\infty\big) = 1 - r.$$

So, any coalgebra defines a cone over (23). We have shown in Lemma 6 that any *exchangeable* coalgebra defines a cone over the draw-delete chain (6). We use this to show that the bernoulli coalgebra *bern*: $[0,1] \to 2 \times [0,1]$ is the final exchangeable coalgebra.

The final coalgebra $2^\infty \to 2 \times 2^\infty$ is not exchangeable. To see this, consider the infinite alternating sequence $01010101 \cdots \in 2^\infty$, so that the left hand side of (14) at the sequence gives $(0, 1, 010101\ldots)$, whereas the right hand side of (14) at the sequence gives $(1, 0, 010101\ldots)$, which is different.

The bernoulli coalgebra *bern*: $[0,1] \to 2 \times [0,1]$ *is* exchangeable, because

$$(\text{id} \otimes bern) = \!\!\ll bern(r) = r^2 \big| 1,1,r \big\rangle$$
$$+ (1-r)r \big| 0,1,r \big\rangle + r(1-r) \big| 1,0,r \big\rangle$$
$$+ (1-r)^2 \big| 0,0,r \big\rangle.$$

Notice that reordering the outputs does not change the statistics.

Theorem 12. *The bernoulli coalgebra bern:* $[0,1] \to 2 \times [0,1]$ *is the final exchangeable coalgebra.*

Proof. Let $h\colon X \to 2 \times X$ be an exchangeable coalgebra. By Lemma 6 the associated maps $h_K\colon X \to \mathcal{M}[K](2)$ form a cone. By de Finetti's theorem in limit form—that is, by Theorem 8—we get a unique mediating channel $f\colon X \to [0,1]$ with $binom[K] \circ f = h_K$ for each K. Our aim is to show that f is a unique coalgebra map: $bern \circ f = (\text{id} \otimes f) \circ h$.

We do know that the limiting map $X \xrightarrow{\ f\ } [0,1] \xrightarrow{\ iid\ } 2^\infty$ is a unique coalgebra homomorphism. So to show that f is a unique coalgebra homomorphism, it suffices to show that *iid* is a monomorphism in $\mathcal{K}\ell(\mathcal{G})$.

To see this, first notice that the K-th moment of some distribution $\omega \in \mathcal{G}([0,1])$ is the probability $(iid =\!\!\ll \omega)(\{1\}^K \times 2^\infty)$, and so if two random sequences are equal, $iid =\!\!\ll \omega_1 = iid =\!\!\ll \omega_2$, then in particular the moments of ω_1 and ω_2 are equal and so $\omega_1 = \omega_2$. Moreover the endofunctor $2\times -$ preserves monomorphisms. So if the final channel $X \to 2^\infty$ factors through $iid : [0,1] \to 2^\infty$ as a channel then it also factors through as a coalgebra homomorphism, necessarily uniquely, by the definition of monomorphism. $\qquad\square$

Example 13. We return to the example of Pólya's urn coalgebra *pol*: $\mathcal{M}_*(2) \to 2 \times \mathcal{M}_*(2)$. We have shown in Proposition 4 that the cone $pol_K : \mathcal{M}_*(2) \to \mathcal{M}[K](2)$ defined by iterated draws is given by the beta-bernoulli distributions: $pol_K(\varphi) = betabern[K](\varphi(0), \varphi(1))$, and we have shown in Proposition 7 that the beta-bernoulli cone over (6) has mediating map the beta distribution. So the final channel $beta : \mathcal{M}_*(2) \to [0,1]$ is given by putting $beta(\varphi)$ as the beta distribution with parameters $(\varphi(0), \varphi(1))$, and this is a coalgebra homomorphism. The commuting diagram

$$
\begin{array}{ccc}
\mathcal{M}_*(2) & \xrightarrow{\ pol\ } & 2 \times \mathcal{M}_*(2) \\
{\scriptstyle beta}\downarrow & & \downarrow{\scriptstyle 2\times beta} \\
[0,1] & \xrightarrow[\ bern\ ]{} & 2 \times [0,1]
\end{array}
$$

expresses the 'conjugacy' between the bernoulli and beta distributions, see also [13].

8   Conclusions and Further Work

This paper has translated the classical representation theorem of de Finetti into the language of multisets and distributions and turned it into a limit theorem. This limit is subsequently used for a finality result for exchangeable coalgebras.

We have concentrated, as is usual for the de Finetti theorem, on the binary case, giving a limit statement for the special case $X = 2 = \{0,1\}$ of the cone in Proposition 2. There is an obvious question whether this formulation can be extended to $X = \{0,1,\ldots,n-1\}$ for $n > 2$, or to more general measurable spaces. We see two possible avenues:

- extension to finite sets via a multivariate version of Hausdorff's moments theorem, see [15];
- extension to Polish spaces, following [2] and its categorical justification for extending de Finetti's theorem to such spaces (esp. Theorems 35 and 40) in terms of exchangeability.

Another point for future work is to investigate whether the categorical formulation of de Finetti's theorem could be taken as an axiom for a 'synthetic probability theory' (*e.g.* following [7,8,17]), and indeed whether it holds in some proposed categorical models which do support de-Finetti-like arguments [11,18].

Finally, we note that chains of channels between multisets also arise in recent exponential modalities for linear logic [3,9]; the connections remain to be worked out.

Acknowledgements. Staton is supported by a Royal Society Fellowship, and has enjoyed discussions about formulations of de Finetti's theorem with many people including coauthors on [18].

References

1. Ackerman, N., Freer, C., Roy, D.: Exchangeable random primitives. In: Proceedings of PPS 2016 (2016)
2. Dahlqvist, F., Danos, V., Garnier, I.: Robustly parameterised higher-order probabilistic models. In: Desharnais, J., Jagadeesan, R. (eds.) CONCUR 2016 – Concurrency Theory. LIPIcs, vol. 59, pp. 23:1–23:15. Schloss Dagstuhl (2016)
3. Dahlqvist, F., Kozen, D.: Semantics of higher-order probabilistic programs with conditioning. In: Principles of Programming Languages, pp. 57:1–57:29. ACM Press (2020). https://doi.org/10.1145/3371125
4. Danos, V., Garnier, I.: Dirichlet is natural. In: Ghica, D. (ed.) Mathematical Foundations of Programming Semantics, number 319 in Electronic Notes in Theoretical Computer Science, pp. 137–164. Elsevier, Amsterdam (2015)

5. de Finetti, B.: Funzione caratteristica di un fenomeno aleatorio. Memorie della R. Accademia Nazionale dei Lincei, IV, fasc. **5**, 86–113 (1930). www.brunodefinetti. it/Opere/funzioneCaratteristica.pdf
6. Feller, W.: An Introduction to Probability Theory and Its applications, vol. II. Wiley, Hoboken (1970)
7. Fritz, T.: A synthetic approach to Markov kernels, conditional independence, and theorems on sufficient statistics. arxiv.org/abs/1908.07021 (2019)
8. Fritz, T., Rischel, E.: The zero-one laws of Kolmogorov and Hewitt-Savage in categorical probability. arxiv.org/abs/1912.02769 (2019)
9. Hamano, M.: A linear exponential comonad in s-finite transition kernels and probabilistic coherent spaces. arxiv:1909.07589, September 2019
10. Hausdorff, F.: Summationsmethoden und Momentfolgen I. Math. Zeitschr. **9**, 74–109 (1921)
11. Heunen, C., Kammar, O., Staton, S., Yang, H.: A convenient category for higher-order probability theory. In: Logic in Computer Science, pp. 1–12. IEEE Computer Society (2017)
12. Jacobs, B.: Structured probabilistic reasoning (2019, forthcoming book). http:// www.cs.ru.nl/B.Jacobs/PAPERS/ProbabilisticReasoning.pdf
13. Jacobs, B.: A channel-based perspective on conjugate priors. Math. Struct. Comput. Sci. **30**(1), 44–61 (2020)
14. Kerstan, H., König, B.: Coalgebraic trace semantics for probabilistic transition systems based on measure theory. In: Koutny, M., Ulidowski, I. (eds.) CONCUR 2012. LNCS, vol. 7454, pp. 410–424. Springer, Heidelberg (2012). https://doi.org/ 10.1007/978-3-642-32940-1_29
15. Kleiber, C., Stoyanov, J.: Multivariate distributions and the moment problem. J. Multivar. Anal. **113**, 7–18 (2013)
16. Klenke, A.: Probability Theory. Springer, London (2013). https://doi.org/10.1007/ 978-1-4471-5361-0
17. Ścibior, A., et al.: Denotational validation of higher-order Bayesian inference. In: Principles of Programming Languages, pp. 60:1–60:29. ACM Press (2018)
18. Staton, S., Stein, D., Yang, H., Ackerman, N., Freer, C., Roy, D.: The Beta-Bernoulli process and algebraic effects. In: Chatzigiannakis, I., Kaklamanis, C., Marx, D., Sannella, D. (eds.) International Colloquium on Automata, Languages and Programming. LIPIcs, vol. 107, pp. 141:1–141:15. Schloss Dagstuhl (2018)

Injective Objects and Fibered Codensity Liftings

Yuichi Komorida[1,2(✉)]

[1] The Graduate University for Advanced Studies, SOKENDAI, Tokyo, Japan
[2] National Institute of Informatics, Tokyo, Japan
komorin@nii.ac.jp

Abstract. Functor lifting along a fibration is used for several different purposes in computer science. In the theory of coalgebras, it is used to define coinductive predicates, such as simulation preorder and bisimilarity. Codensity lifting is a scheme to obtain a functor lifting along a fibration. It generalizes a few previous lifting schemes including the Kantorovich lifting. In this paper, we seek a property of functor lifting called fiberedness. Hinted by a known result for Kantorovich lifting, we identify a sufficient condition for a codensity lifting to be fibered. We see that this condition applies to many examples that have been studied. As an application, we derive some results on bisimilarity-like notions.

1 Introduction

In this paper, we focus on a category-theoretical gadget, called *functor lifting*, and seek a property thereof, called *fiberedness*. As is often the case with such mathematical topics, functor lifting comes up in several different places in computer science under various disguises (as mentioned in Sect. 1.6). Here we see one of such places, *bisimilarity and its generalizations on coalgebras*, before we formally introduce functor lifting.

1.1 Coalgebras and Bisimilarity

Computer programs work as we write them, not necessarily as we expect. One approach to overcome this gap is to *verify* the systems so that we can make sure that they meet our requirements. Abstract mathematical methods are often useful for the purpose, but before that, we have to *model* the target system by some mathematical structure.

Coalgebra [27] is one of such mathematical structure with a broad scope of application. It is defined in terms of the theory of *categories and functors*. Given a category $\mathbb{C}$ and an endofunctor $F\colon \mathbb{C} \to \mathbb{C}$, an F-coalgebra is defined

---

The author was supported by ERATO HASUO Metamathematics for Systems Design Project (No. JPMJER1603), JST.

D. Petrişan and J. Rot (Eds.): CMCS 2020, LNCS 12094, pp. 112–132, 2020.
https://doi.org/10.1007/978-3-030-57201-3_7

as an arrow $c\colon X \to FX$. This simple definition includes many kinds of state-transition systems as special cases, e.g., Kripke frame (and model), Markov chain (and process), and (deterministic and non-deterministic) automata.

Having modeled a system as a coalgebra, we can ask a fundamental question: which states behave the same? *Bisimilarity* [25,26] is one of the notions to define such equivalence. (For an introduction, see, e.g., [28].) We sketch the idea in the case where $\mathbb{C} = \mathbf{Set}$ and $F = \Sigma \times (-)$. In this case, F-coalgebras are deterministic LTSs. Consider a coalgebra $c\colon X \to \Sigma \times X$ and define $l\colon X \to \Sigma$ and $n\colon X \to X$ by $(l(x), n(x)) = c(x)$. The point here is the following observation: if $x, y \in X$ behave the same, then $l(x) = l(y)$ must hold, and $n(x)$ and $n(y)$ must behave the same. This is almost the definition of bisimilarity: the bisimilarity relation is the greatest binary relation $\sim \subseteq X \times X$ that satisfies

$$x \sim y \implies l(x) = l(y) \wedge n(x) \sim n(y).$$

For other functors F, the idea is roughly the same: in a coalgebra $c\colon X \to FX$, for $x, y \in X$ to behave the same, $c(x)$ and $c(y)$ must behave the same. To define bisimilarity precisely, however, we have to turn a relation $R \subseteq X \times X$ into $R' \subseteq FX \times FX$.

1.2 Qualitative and Quantitative Bisimilarity from Functor Lifting

An elegant way to formulate this is the following: bundle binary relations on all sets into one *fibration* and use *functor lifting* as in [11]. We give ideas on them here. The precise definitions are in Sect. 2.

First, we gather all pairs (X, R) of a set X and a binary relation $R \subseteq X \times X$ into one category $\mathbf{ERel}$ (Example 8). It comes with a forgetful functor $U\colon \mathbf{ERel} \to \mathbf{Set}$. (This is a *fibration*.) Any binary relation R on X is sent to X by U; placing the things vertically, R is "above" X. Now let us assume that there exists a functor $\dot{F}\colon \mathbf{ERel} \to \mathbf{ERel}$ satisfying $U \circ \dot{F} = F \circ U$. This means that any binary relation R on X is sent to one on FX:

$$
\begin{array}{ccc}
\mathbf{ERel} \xrightarrow{\ \dot{F}\ } \mathbf{ERel} & \qquad & R \longmapsto \dot{F}R \\
\ \ \downarrow{\scriptstyle U} \qquad\quad \downarrow{\scriptstyle U} & & \ \vdots \qquad\quad \vdots \\
\mathbf{Set} \xrightarrow[\ F\ ]{} \mathbf{Set} & & X \longmapsto FX
\end{array}
$$

(This means that the functor $\dot{F}$ is a *lifting* of F along U.) The functor $U\colon \mathbf{ERel} \to \mathbf{Set}$ has an important structure: for any $f\colon Y \to X$ and a relation R on X, we can obtain a relation $f^{*}R$ on Y in a canonical way:

$$f^{*}R = \{(y, y') \in Y \times Y \,|\, (f(y), f(y')) \in R\}.$$

(This is called *reindexing* or *pullback*.) By using these, we can define the bisimulation relation on $c\colon X \to FX$ as the greatest fixed point of $f^{*} \circ \dot{F}$.

An advantage of this approach is that we can readily generalize this to other "bisimilarity-like" notions. For example, by changing the fibration to $\mathbf{PMet}_\top \to \mathbf{Set}$ (Example 7), one can define a *behavioral (pseudo)metric* [2].

1.3 Codensity Lifting of Endofunctors

Now we know that a functor lifting induces a bisimilarity-like notion. Then, how can we obtain a functor lifting? *Codensity Lifting* is a scheme to obtain such liftings. It is first introduced by Katsumata and Sato [16] for monads using *codensity monad* construction [22]. It is later extended to general endofunctors by Sprunger et al. [31]. It is parametrized in a set of data called a *lifting parameter*. By changing lifting parameters, a broad class of functor liftings can be represented as codensity liftings, as is shown, e.g., in Komorida et al. [18].

As mentioned in the last section, we can define a bisimilarity-like notion using codensity lifting. It is called *codensity bisimilarity* in [18, Sections III and VI].

1.4 Fiberedness of Lifting

In some situations, we have to assume that $\dot{F}\colon \mathbb{E} \to \mathbb{E}$ interacts well with the pullback operation between the fibers. In such a situation, $\dot{F}$ is required to be *fibered* (Definition 11). It means that pullbacks and $\dot{F}$ are "commutative," in the sense that they satisfy $\dot{F}(f^*P) = (Ff)^*(\dot{F}P)$.

Some of the existing works indeed require fiberedness. For example, Hasuo et al. [9, Definition 2.2] include fiberedness in their definition of predicate lifting. Fiberedness also plays a notable role in [3], where it is rephrased to isometry-preservation. However, there has been no systematic result on fiberedness of codensity lifting.

1.5 Contributions

In the current paper, hinted by a result of Baldan et al. [3], we show a sufficient condition on the lifting parameter guaranteeing the resulting functor to be fibered (Theorem 20). The scope of our fiberedness result is so broad that it covers, e.g., most of the examples presented in [18] (Sect. 5).

The condition involves a variation of the notion of injective object, which we call *c-injective object* (Definition 15). To our knowledge, such a notion connecting injective objects and fibrations is new. We study some basic properties of them.

Using the fiberedness result, we show a property of codensity bisimilarity which we call *stability under coalgebra morphisms* (Proposition 49). As a corollary, we see that, when there is a final coalgebra, the codensity bisimilarity on any coalgebra is determined by that on the final coalgebra. Note that this kind of property is well-known for a conventional bisimilarity relation (Corollary 50).

To summarize, our technical contributions are as follows:

- We define *c-injective objects* for fibrations (Definition 15) and show some properties of them.
- We show a sufficient condition on the lifting parameter to guarantee fiberedness of codensity lifting (Theorem 20 and Corollary 24).
- We show a number of examples (Sect. 5) to which the condition above is applicable.

– As an application, we show that codensity bisimilarity is stable under coalgebra morphisms (Proposition 49) in many cases, including a new one (Example 51).

1.6   Related Work

Even though we focused on bisimilarity and coalgebra above, functor lifting comes up in computer science here and there. To name a few, it has applications in logical predicates [11,15], quantitative bisimulation [3], and differential privacy [29].

As mentioned above, there have been many methods to obtain liftings of functors. *Kantorovich lifting* [2,19] and *generalized Kantorovich metric* [5] are both special cases of the version of codensity lifting considered here. *Categorical* $\top\top$-*lifting* [15] is the precursor of the original version of codensity lifting, but it is not a special case of codensity lifting. For categorical $\top\top$-lifting, one uses internal Hom-objects rather than Hom-sets like codensity lifting. Obtaining a sufficient condition for fiberedness of categorical $\top\top$-lifting is future work. *Wasserstein lifting* [2] is another method that is somehow dual to Kantorovich lifting. They have shown that any lifting obtained by this scheme is fibered. Klin [17] goes a different way: rather than showing fiberedness, they incorporate fiberedness in the definition. They study *the least fibered lifting* along **EqRel** → **Set** and show that, in good situations, it coincides with the *canonical relation lifting*.

The notion of *injective object* is first introduced in homological algebra as *injective modules* [1]. There are also some works about injective objects outside homological algebra: Scott [30] and Banaschewski and Bruns [4] have identified the injective objects in **Top₀** and **Pos**, respectively. We use their results in Sect. 4 (where the categories mentioned are defined). Injective objects w.r.t. isometric embeddings in the category of metric spaces are also well-studied and called *hyperconvex spaces* [7]. Finding a precise connection between them and c-injective objects in **PMet⊤** → **Set** (Example 7) is future work. Recently, in his preprint [8], Fujii has extended the above result in **Pos** and characterized injective objects in the category of $\mathcal{Q}$-categories with respect to the class of fully faithful $\mathcal{Q}$-functors, for any quantale $\mathcal{Q}$.

1.7   Organization

In Sect. 2, we review **CLat**$_\sqcap$-*fibrations* and *functor liftings*. In Sect. 3, we review the definition of *codensity lifting* and introduce the notion of *c-injective objects*. We show a sufficient condition for a codensity lifting to be fibered. In Sect. 4, we show some general results on c-injective objects. In Sect. 5, we list several examples of fibered codensity liftings using the results in Sect. 4. In Sect. 6, we apply the fiberedness result to *codensity bisimilarity*. In Sect. 7, we conclude with some remarks and future work.

2   Preliminaries

We assume some knowledge of *category theory*, but the full content of the standard reference [24] is not needed. The basic definitions and theorems, e.g., those in Leinster [23], is enough. Even though we have explained our motivation through coalgebra, no knowledge of coalgebra is needed for the main result in Sect. 3.

In the following, **Set** means the category of sets and (set-theoretic) functions.

2.1   CLat$_\sqcap$-Fibrations

Here we introduce **CLat**$_\sqcap$-*fibrations*, as defined in [18]. We use them to model various "notions of indistinguishability" like preorder, equivalence relation, and pseudometric. Assuming full knowledge of the theory of fibrations, we could define them as poset fibrations with fibered small meets. Instead, we give an explicit definition below. This is mainly because we need the notion of *Cartesian arrow*. For a comprehensive account of the theory of fibrations, the reader can consult, e.g., a book by Jacobs [13] or Hermida's thesis [10], but in the following, we do not assume any knowledge of fibrations.

We first define a fiber of a functor over an object. Basically, this is only considered in the case where the functor is a fibration.

Definition 1 (fiber). Let $p\colon \mathbb{E} \to \mathbb{C}$ be a functor and $X \in \mathbb{C}$ be an object. The *fiber over* X is the subcategory of $\mathbb{E}$

- whose objects are $P \in \mathbb{E}$ such that $pP = X$ and
- whose arrows are $f\colon P \to Q$ such that $pf = \mathrm{id}_X$.

We denote it by $\mathbb{E}_X$.

Note that, if p is faithful, then each fiber is a thin category, i.e., a preorder. The following definition of poset fibration is a special case of that in [13].

Definition 2 (cartesian arrow and poset fibration). Let $p\colon \mathbb{E} \to \mathbb{C}$ be a faithful functor.

An arrow $f\colon P \to Q$ in $\mathbb{E}$ is *Cartesian* if the following condition is satisfied:

- For each $R \in \mathbb{E}$ and $g\colon R \to Q$, there exists $h\colon R \to P$ such that $g = f \circ h$ if and only if there exists $h'\colon pR \to pP$ such that $pg = pf \circ h'$.

The functor p is called a *poset fibration* if the following are satisfied:

- For each $X \in \mathbb{C}$, the fiber $\mathbb{E}_X$ is a poset. The order is denoted by $\sqsubseteq$. We define the direction so that $P \sqsubseteq Q$ holds if and only if there is an arrow $P \to Q$ in $\mathbb{E}_X$.
- For each $Q \in \mathbb{E}$ and $f\colon X \to pQ$, there exists an object $f^*Q \in \mathbb{E}_X$ and a Cartesian arrow $\dot{f}\colon f^*Q \to Q$ such that $p\dot{f} = f$. (Such f^*Q and $\dot{f}$ are necessarily unique.)

The map $Q \mapsto f^*Q$ turns out to be a monotone map from $\mathbb{E}_Y$ to $\mathbb{E}_X$. We call it the *pullback functor* and denote it by $f^* \colon \mathbb{E}_Y \to \mathbb{E}_X$.

Intuitively, pullback functors model substitutions. Indeed, in many examples, they are just "assigning $f(x)$ to y", as can be seen below.

Example 3 (pseudometric). Let $\top$ be a positive real number or $+\infty$. Define a category $\mathbf{PMet}_\top$ as follows:

- Each object is a pair (X, d) of a set X and a $[0, \top]$-valued pseudometric $d \colon X \times X \to [0, \top]$. (A pseudometric is a metric without the condition $d(x, y) = 0 \implies x = y$.)
- Each arrow from (X, d_X) to (Y, d_Y) is a nonexpansive map $f \colon X \to Y$. (f is nonexpansive if, for all x and $x' \in X$, $d_X(x, x') \geq d_Y(f(x), f(x'))$.)

The obvious forgetful functor $\mathbf{PMet}_\top \to \mathbf{Set}$ is a poset fibration. For each $X \in \mathbf{Set}$, the objects of the fiber $(\mathbf{PMet}_\top)_X$ are the pseudometrics on X. However, the order is reversed: in our notation, the order is defined by

$$(X, d_1) \sqsubseteq (X, d_2) \Leftrightarrow \forall x, x' \in X, d_1(x, x') \geq d_2(x, x').$$

An arrow $f \colon (X, d_X) \to (Y, d_Y)$ is Cartesian if and only if it is an isometry, i.e., $d_X(x, x') = d_Y(f(x), f(x'))$ holds for all x, x'. For $(Y, d_Y) \in \mathbf{PMet}_\top$ and $f \colon X \to Y$, the pullback $f^*(Y, d_Y)$ is the set X with the pseudometric $(x, x') \mapsto d_Y(f(x), f(x'))$.

We list a few properties of pullback functors that we use:

Proposition 4. Let $p \colon \mathbb{E} \to \mathbb{C}$ be a poset fibration, $f \colon X \to Y$ be an arrow in $\mathbb{C}$ and $P \in \mathbb{E}_X$ and $Q \in \mathbb{E}_Y$ be objects in $\mathbb{E}$. There exists an arrow $g \colon P \to Q$ such that $pg = f$ if and only if $P \sqsubseteq f^*Q$. Moreover, such g is Cartesian if and only if $P = f^*Q$. □

Proposition 5. Let $p \colon \mathbb{E} \to \mathbb{C}$ be a poset fibration.

- For each $X \in \mathbb{C}$, $(\mathrm{id}_X)^* \colon \mathbb{E}_X \to \mathbb{E}_X$ is the identity functor.
- For each composable pair of arrows $X \xrightarrow{f} Y \xrightarrow{g} Z$ in $\mathbb{C}$, $(g \circ f)^* = f^* \circ g^*$ holds. □

Now we define the class that we are concerned about, $\mathbf{CLat}_\sqcap$-fibrations.

Definition 6 ($\mathbf{CLat}_\sqcap$-fibration). A poset fibration $p \colon \mathbb{E} \to \mathbb{C}$ is a $\mathbf{CLat}_\sqcap$-*fibration* if the following conditions are satisfied:

- Each fiber $\mathbb{E}_X$ is small and has small meets, which we denote by $\sqcap$.
- Each pullback functor f^* preserves small meets.

Note that, in the situation above, each fiber $\mathbb{E}_X$ is a complete lattice: the small joins can be constructed using small meets.

Example 7 (pseudometric). The poset fibration $\mathbf{PMet}_\top \to \mathbf{Set}$ in Example 3 is a $\mathbf{CLat}_\sqcap$-fibration. Indeed, meets can be defined by sups of pseudometrics: if we let $(X, d) = \sqcap_{a \in A}(X, d_a)$, then

$$d(x, x') = \sup_{a \in A} d_a(x, x')$$

holds.

Example 8 (binary relations). Define a category $\mathbf{ERel}$ of sets with an endorelation as follows:

- Each object is a pair (X, R) of a set X and a binary relation $R \subseteq X \times X$.
- Each arrow from (X, R_X) to (Y, R_Y) is a map $f \colon X \to Y$ preserving the relations; that is, we require f to satisfy $(x, x') \in R_X \implies (f(x), f(x')) \in R_Y$.

The obvious forgetful functor $\mathbf{ERel} \to \mathbf{Set}$ is a $\mathbf{CLat}_\sqcap$-fibration. For each $X \in \mathbf{Set}$, the fiber $\mathbf{ERel}_X$ is the complete lattice of subsets of $X \times X$.

An arrow $f \colon (X, R_X) \to (Y, R_Y)$ is Cartesian if and only if it reflects the relations, i.e., $(x, x') \in R_X \Leftrightarrow (f(x), f(x')) \in R_Y$ holds for all x, x'. For $(Y, R_Y) \in \mathbf{ERel}$ and $f \colon X \to Y$, the pullback $f^*(Y, R_Y)$ is the set X with the relation $\{(x, x') \in X \times X | (f(x), f(x')) \in R_Y\}$.

Define the following full subcategories of $\mathbf{ERel}$:

- The category $\mathbf{Pre}$ of preordered sets and monotone maps.
- The category $\mathbf{EqRel}$ of sets with an equivalence relation and maps preserving them.

The forgetful functors $\mathbf{Pre} \to \mathbf{Set}$ and $\mathbf{EqRel} \to \mathbf{Set}$ are also $\mathbf{CLat}_\sqcap$-fibrations.

$\mathbf{CLat}_\sqcap$-fibrations are not necessarily "relation-like". There also is an example with a much more "space-like" flavor.

Example 9. The forgetful functor $\mathbf{Top} \to \mathbf{Set}$ from the category $\mathbf{Top}$ of topological spaces and continuous maps is a $\mathbf{CLat}_\sqcap$-fibration.

2.2 Lifting and Fiberedness

Another pivotal notion in the current paper is *functor lifting*. In Sect. 1.2 we have seen that it is used to define bisimilarity, or more generally bisimilarity-like notions, as a way to turn a relation (or pseudometric, etc.) on X into one on FX. Here we review the formal definition in a restricted form that only considers $\mathbf{CLat}_\sqcap$-fibration. (Note that, usually it is defined more generally, and there are indeed applications of such general definition.)

Definition 10 (lifting of endofunctor). Let $p \colon \mathbb{E} \to \mathbb{C}$ be a $\mathbf{CLat}_\sqcap$-fibration and $F \colon \mathbb{C} \to \mathbb{C}$ be a functor. A *lifting* of F along p is a functor $\dot{F} \colon \mathbb{E} \to \mathbb{E}$ such that $p \circ \dot{F} = F \circ p$ holds:

$$\begin{array}{ccc}
\mathbb{E} & \xrightarrow{\dot{F}} & \mathbb{E} \\
{\scriptstyle p}\downarrow & & \downarrow{\scriptstyle p} \\
\mathbb{C} & \xrightarrow{F} & \mathbb{C}.
\end{array}$$

We then define *fiberedness* of a lifting. This means that the lifting interacts well with the pullback structure of the fibration, but we first give a definition focusing on Cartesian arrows. Here we define it in a slightly more general way so that we can use them later (Sect. 4).

Definition 11 (fibered functor [13, Definition 1.7.1]**).** Let $p\colon \mathbb{E} \to \mathbb{C}$ and $q\colon \mathbb{F} \to \mathbb{D}$ be $\mathbf{CLat}_\sqcap$-fibrations. A *fibered functor* from p to q is a functor $\dot{F}\colon \mathbb{E} \to \mathbb{F}$ such that there is another functor $F\colon \mathbb{C} \to \mathbb{D}$ satisfying $q \circ \dot{F} = F \circ p$ and $\dot{F}$ sends each Cartesian arrow to a Cartesian arrow.

Note that, in the situation above, such F is uniquely determined by p, q, and $\dot{F}$.

Now we see a characterization of fiberedness by means of pullback.

Proposition 12. Let $p\colon \mathbb{E} \to \mathbb{C}$ and $q\colon \mathbb{F} \to \mathbb{D}$ be $\mathbf{CLat}_\sqcap$-fibrations and $\dot{F}\colon \mathbb{E} \to \mathbb{F}$ and $F\colon \mathbb{C} \to \mathbb{D}$ be functors satisfying $q \circ \dot{F} = F \circ p$. $\dot{F}$ is a fibered functor if and only if, for any $f\colon X \to Y$ in $\mathbb{C}$ and $P \in \mathbb{E}_Y$, $\dot{F}(f^*P) = (Ff)^*(\dot{F}P)$ holds. $\qquad\square$

We use this in the proof of the main result.

3   C-Injective Objects and Codensity Lifting

3.1   Codensity Lifting

Before we formulate our main result, we introduce *codensity lifting* of endofunctors [16,31]. Here we use an explicit definition for a narrower situation than the original one.

Definition 13 (codensity lifting (as in [18]**)).** Let

- $p\colon \mathbb{E} \to \mathbb{C}$ be a $\mathbf{CLat}_\sqcap$-fibration,
- $F\colon \mathbb{C} \to \mathbb{C}$ be a functor,
- $\Omega \in \mathbb{E}$ be an object above $\Omega \in \mathbb{C}$, and
- $\tau\colon F\Omega \to \Omega$ be an F-algebra.

Define a functor $F^{\Omega,\tau}\colon \mathbb{E} \to \mathbb{E}$, which is a lifting of F along p, by

$$F^{\Omega,\tau}P = \bigsqcap_{f\in\mathbb{E}(P,\Omega)} (F(pf))^*\tau^*\Omega$$

for each $P \in \mathbb{E}$. The functor $F^{\Omega,\tau}$ is called a *codensity lifting* of F. Note that, for each $P \in \mathbb{E}$ and $f \colon P \to \Omega$, the situation is as follows:

$$
\begin{array}{ccc}
& & \Omega \\
& & \vdots \\
FpP \xrightarrow{\ F(pf)\ } F\Omega & \xrightarrow{\ \tau\ } & \Omega
\end{array}
$$

and we can indeed obtain the pullback $(F(pf))^*\tau^*\Omega$.

We have given only the object part of $F^{\Omega,\tau}$ above, but the arrow part, if it is well-defined, should be determined uniquely since p is faithful. We give a proof that it is indeed well-defined. For each $f \colon P \to Q$, we need another arrow $g \colon F^{\Omega,\tau}P \to F^{\Omega,\tau}Q$ such that $pg = F(pf)$. By Proposition 4, it suffices to show the following proposition:

Proposition 14. For any $f \colon P \to Q$, $F^{\Omega,\tau}P \sqsubseteq (F(pf))^* \left(F^{\Omega,\tau}Q \right)$ holds.

Proof. By definition, the l.h.s. satisfies

$$
F^{\Omega,\tau}P = \bigsqcap_{g \in \mathbb{E}(P,\Omega)} (F(pg))^*\tau^*\Omega.
$$

On the other hand, the r.h.s. satisfies

$$
\begin{aligned}
(F(pf))^* \left(F^{\Omega,\tau}Q \right) &= (F(pf))^* \left(\bigsqcap_{h \in \mathbb{E}(Q,\Omega)} (F(ph))^*\tau^*\Omega \right) \\
&= \bigsqcap_{h \in \mathbb{E}(Q,\Omega)} (F(pf))^*(F(ph))^*\tau^*\Omega \\
&= \bigsqcap_{h \in \mathbb{E}(Q,\Omega)} (F(p(h \circ f)))^*\tau^*\Omega.
\end{aligned}
$$

Here, since $\{g \in \mathbb{E}(P,\Omega)\} \supseteq \{h \circ f \mid h \in \mathbb{E}(Q,\Omega)\}$ holds, we have

$$
\bigsqcap_{g \in \mathbb{E}(P,\Omega)} (F(pg))^*\tau^*\Omega \sqsubseteq \bigsqcap_{h \in \mathbb{E}(Q,\Omega)} (F(p(h \circ f)))^*\tau^*\Omega.
$$

This means $F^{\Omega,\tau}P \sqsubseteq (F(pf))^* \left(F^{\Omega,\tau}Q \right)$.   □

3.2   C-Injective Object

In the proof of the functoriality of $F^{\Omega,\tau}$, ultimately we use the fact that, for any $f \colon P \to Q$, any "test" $k \colon Q \to \Omega$ can be turned into another "test" $k \circ f \colon P \to \Omega$. On the other hand, when we try to prove fiberedness of $F^{\Omega,\tau}$, we have to somehow lift a "test" $g \colon P \to \Omega$ along a Cartesian arrow $f \colon P \to Q$ and obtain another "test" $h \colon Q \to \Omega$. This observation leads us to the following definition of *c-injective object*. (The letter c here comes from *Cartesian*.)

Definition 15 (c-injective object). Let $p\colon \mathbb{E} \to \mathbb{C}$ be a fibration. An object $\Omega \in \mathbb{E}$ is a *c-injective object* if the functor $\mathbb{E}(-, \Omega)\colon \mathbb{E}^{\mathrm{op}} \to \mathbf{Set}$ sends every Cartesian arrow to a surjective map.

Equivalently, $\Omega \in \mathbb{E}$ is a c-injective object if, for any Cartesian arrow $f\colon P \to Q$ in $\mathbb{E}$ and any (not necessarily Cartesian) arrow $g\colon P \to \Omega$, there is a (not necessarily Cartesian) arrow $h\colon Q \to \Omega$ satisfying $g = h \circ f$.

Some basic objects can be shown to be c-injective objects.

Example 16 (the two-point set). In the fibration $\mathbf{EqRel} \to \mathbf{Set}$, $(2, =)$ is a c-injective object. Here, $2 = \{\bot, \top\}$ is the two-point set and $=$ means the equality relation. Indeed, for any Cartesian $f\colon (X, R_X) \to (Y, R_Y)$ and any $g\colon (X, R_X) \to (2, =)$, if we define $h\colon (Y, R_Y) \to (2, =)$ by

$$h(y) = \begin{cases} g(x) & \text{if } (y, f(x)) \in R_Y \\ \top & \text{otherwise,} \end{cases}$$

then this turns out to be well-defined and satisfies $h \circ f = g$.

Example 17 (the two-point poset of truth values). In the fibration $\mathbf{Pre} \to \mathbf{Set}$, $(2, \leq)$ is a c-injective object. Here, $\leq$ is the unique partial order satisfying $\bot \leq \top$ and $\top \not\leq \bot$. Indeed, for any Cartesian arrow $f\colon (X, R_X) \to (Y, R_Y)$ and any $g\colon (X, R_X) \to (2, \leq)$, if we define $h\colon\colon (Y, R_Y) \to (2, \leq)$ by

$$h(y) = \begin{cases} \bot & \text{if } (y, f(x)) \in R_Y \text{ for some } x \text{ such that } g(x) = \bot \\ \top & \text{otherwise,} \end{cases}$$

then this turns out to be well-defined and satisfies $h \circ f = g$.

Example 18 (the unit interval as a pseudometric space [3, Theorem 5.8]). In the fibration $\mathbf{PMet}_\top \to \mathbf{Set}$, $[0, \top]$ is a c-injective object. Indeed, for any arrow $g\colon (X, d_X) \to ([0, \top], d_e)$ and any Cartesian arrow $f\colon (X, d_X) \to (Y, d_Y)$, we can show that the map $h\colon Y \to [0, \top]$ defined by $h(y) = \inf_{x \in X} (g(x) + d_Y(f(x), y))$ is nonexpansive from (Y, d_Y) to $([0, \top], d_e)$.

The following non-example shows that c-injectivity crucially depends on the fibration we consider.

Example 19 (non-example). In contrast to Example 17, in the fibration $\mathbf{ERel} \to \mathbf{Set}$, $(2, \leq)$ is not c-injective, where $2 = \{\bot, \top\}$ is the two-point set and $\leq$ is the unique partial order satisfying $\bot \leq \top$ and $\top \not\leq \bot$.

This can be seen as follows. Let $X = \{a, b\}$, $Y = \{x, y, z\}$, $R_X = \emptyset$, and $R_Y = \{(x, z), (z, y)\}$. Then (X, R_X) and (Y, R_Y) are objects of $\mathbf{ERel}$. Consider the maps $f\colon (X, R_X) \to (Y, R_Y)$ and $g\colon (X, R_X) \to (2, \leq)$ defined by $f(a) = x$, $f(b) = y$, $g(a) = \top$, and $g(b) = \bot$. Note that f is Cartesian. However, there is no $h\colon (Y, R_Y) \to (2, \leq)$ such that $h \circ f = g$: such h would satisfy $\top = h(f(a)) = h(x) \leq h(z) \leq h(y) = h(f(b)) = \bot$, which contradict $\top \not\leq \bot$.

The same example can also be used to show that, in contrast to Example 16, $(2, =)$ is not c-injective, where $=$ means the equality relation.

3.3 Sufficient Condition for Fibered Codensity Lifting

Now we are prepared to state the following main theorem of the current paper. The strategy of the proof is roughly as mentioned earlier.

Theorem 20 (fiberedness from injective object). In the setting of Definition 13, if Ω is a c-injective object, then $F^{\Omega,\tau}$ is fibered.

Proof. Let $f\colon P \to Q$ be any Cartesian arrow. By Proposition 12, it suffices to show $F^{\Omega,\tau}P = (F(pf))^* \left(F^{\Omega,\tau}Q\right)$. Here, $F^{\Omega,\tau}P \sqsubseteq (F(pf))^* \left(F^{\Omega,\tau}Q\right)$ has already been proven. Thus, our goal is the inequality $F^{\Omega,\tau}P \sqsupseteq (F(pf))^* \left(F^{\Omega,\tau}Q\right)$.

Here, since Ω is c-injective and f is Cartesian, the following inclusion holds:

$$\{g \in \mathbb{E}(P,\Omega)\} \subseteq \{h \circ f \mid h \in \mathbb{E}(Q,\Omega)\}.$$

By the definition of the meet, we have

$$\prod_{g \in \mathbb{E}(P,\Omega)} (F(pg))^* \tau^* \Omega \sqsupseteq \prod_{h \in \mathbb{E}(Q,\Omega)} (F(p(h \circ f)))^* \tau^* \Omega.$$

By the calculation in the proof of Proposition 14, this implies

$$F^{\Omega,\tau}P \sqsupseteq (F(pf))^* \left(F^{\Omega,\tau}Q\right).$$

$\square$

Remark 21. A refinement of Theorem 20 to an if-and-only-if result seems hard. At least there is a simple counterexample to the most naive version of it: Consider a $\mathbf{CLat}_\sqcap$-fibration $\mathrm{Id}\colon \mathbb{C} \to \mathbb{C}$, an endofunctor $\mathrm{Id}\colon \mathbb{C} \to \mathbb{C}$, an object $C \in \mathbb{C}$, and an arrow $\tau\colon C \to C$. The codensity lifting $\mathrm{Id}^{C,\tau}$ is always equal to Id, which is fibered. However, since any arrow in $\mathbb{C}$ is a Cartesian arrow w.r.t. Id, it is not hard to find an example of C and $\mathbb{C}$ such that C is not c-injective w.r.t. Id.

Example 22 (Kantorovich lifting). Baldan et al. [3, Theorem 5.8] have shown that any Kantorovich lifting preserves isometries. In terms of fibrations, this means that such functor is a fibered endofunctor on the fibration $\mathbf{PMet}_\top \to \mathbf{Set}$.

Since Kantorovich lifting is a special case of codensity lifting where $\Omega = ([0,\top], d_\mathbb{R})$, Theorem 20 and Example 18 recover the same result. Actually, this has inspired Theorem 20 as a prototype.

The argument above also applies to situations with multiple parameters.

Definition 23 (codensity lifting with multiple parameters (as in [18])). Let $\mathbb{E}, \mathbb{C}, p$, and F be as in Definition 13. Let A be a set. Assume that, for each $a \in A$, we are given $\Omega_a \in \mathbb{E}$ above $\varOmega_a \in \mathbb{C}$ and $\tau_a\colon F\varOmega_a \to \varOmega_a$. Define a functor $F^{\Omega,\tau}\colon \mathbb{E} \to \mathbb{E}$ by

$$F^{\Omega,\tau}P = \prod_{a \in A} F^{\Omega_a,\tau_a}P$$

for each $P \in \mathbb{E}$.

Corollary 24. In the setting of Definition 23, if, for each $a \in A$, Ω_a is a c-injective object, then $F^{\Omega,\tau}$ is fibered.

Proof. For any $P \in \mathbb{E}$ above $X \in \mathbb{C}$ and $f : Y \dashrightarrow X$ in $\mathbb{C}$, using Theorem 20, we can see

$$
\begin{aligned}
(Ff)^* F^{\Omega,\tau} P = (Ff)^* \prod_{a \in A} F^{\Omega_a, \tau_a} P \qquad &= \prod_{a \in A} (Ff)^* F^{\Omega_a, \tau_a} P \\
= \prod_{a \in A} F^{\Omega_a, \tau_a} f^* P \qquad &= F^{\Omega,\tau} f^* P.
\end{aligned}
$$
$\square$

Example 25 (Kantorovich lifting with multiple parameters). In [19], König and Mika-Michalski introduced a generalized version of Kantorovich lifting.

Since it is a special case of Definition 23 where p is the fibration $\mathbf{PMet}_\top \to \mathbf{Set}$ and $\Omega = ([0, \top], d_\mathbb{R})$, Corollary 24 and Example 18 imply that such lifting always preserves isometries.

4   Results on C-Injective Objects

Here we seek properties of c-injective objects, mainly to obtain more examples of them. We also see that, in a few fibrations, c-injective objects have been essentially identified by previous works.

4.1 $\mathcal{M}$-injective Objects

To connect c-injectivity with existing works, we consider a more general notion of $\mathcal{M}$-injective object. The following definition is found e.g. in [14, Sect. 9.5].

Definition 26. Let $\mathbb{C}$ be a category and $\mathcal{M}$ be a class of arrows in $\mathbb{C}$. An object $X \in \mathbb{C}$ is an $\mathcal{M}$-*injective object* if the functor $\mathbb{C}(-, X) : \mathbb{C}^{op} \to \mathbf{Set}$ sends every arrow in $\mathcal{M}$ to a surjective map.

The definition of c-injective objects is a special case of the definition above where $\mathcal{M}$ is the class of all Cartesian arrows.

The following is a folklore result. The dual is found e.g. in [12, Proposition 10.2].

Proposition 27. Let $\mathbb{C}, \mathbb{D}$ be categories, $\mathcal{M}_\mathbb{C}, \mathcal{M}_\mathbb{D}$ be classes of arrows, and $L \dashv R : \mathbb{C} \to \mathbb{D}$ be a pair of adjoint functors. Assume that L sends any arrow in $\mathcal{M}_\mathbb{D}$ to one in $\mathcal{M}_\mathbb{C}$. For any $\mathcal{M}_\mathbb{C}$-injective $C \in \mathbb{C}$, $RC \in \mathbb{D}$ is $\mathcal{M}_\mathbb{D}$-injective.

Proof. It suffices to show that $\mathbb{D}(-, RC) : \mathbb{D}^{op} \to \mathbf{Set}$ sends each arrow in $\mathcal{M}_\mathbb{D}$ to a surjective map. By the assumption, the functor above factorizes to $L : \mathbb{D} \to \mathbb{C}$ and $\mathbb{C}(-, C) : \mathbb{D}^{op} \to \mathbf{Set}$. The former sends each arrow in $\mathcal{M}_\mathbb{D}$ to one in $\mathcal{M}_\mathbb{C}$ and the latter sends one in $\mathcal{M}_\mathbb{C}$ to a surjective map. Thus, the composition of these sends each arrow in $\mathcal{M}_\mathbb{D}$ to a surjective map. $\square$

For epireflective subcategories, we have a sharper result:

Proposition 28. In the setting of Proposition 27, assume, in addition,

- R is fully faithful,
- R sends each arrow in $\mathcal{M}_\mathbb{C}$ to one in $\mathcal{M}_\mathbb{D}$, and
- each component of the unit $\eta\colon \mathrm{Id} \to RL$ is an epimorphism in $\mathcal{M}_\mathbb{D}$.

Then, $D \in \mathbb{D}$ is $\mathcal{M}_\mathbb{D}$-injective if and only if it is isomorphic to RC for some $\mathcal{M}_\mathbb{C}$-injective $C \in \mathbb{C}$.

Proof. The "if" part is Proposition 27. We show the "only if" part.

Let $D \in \mathbb{D}$ be any $\mathcal{M}_\mathbb{D}$-injective object. Since $\eta_D\colon D \to RLD$ is in $\mathcal{M}_\mathbb{D}$, we can use the $\mathcal{M}_\mathbb{D}$-injectiveness of D to obtain $f\colon RLD \to D$ such that $f \circ \eta_D = \mathrm{id}_D$. Here, $\eta_D \circ f \circ \eta_D = \eta_D$ and, by epi-ness of η_D, $\eta_D \circ f = \mathrm{id}_{RLD}$. Thus, η_D is an isomorphism.

Now we show that LD is $\mathcal{M}_\mathbb{C}$-injective. Let $f\colon C \to LD$ and $g\colon C \to C'$ be any arrow in $\mathbb{C}$ and assume that g is in $\mathcal{M}_\mathbb{C}$. Send these by R to $\mathbb{D}$ and consider Rf and Rg. By the assumption, Rg is in $\mathcal{M}_\mathbb{D}$. Since RLD is isomorphic to D, it is also $\mathcal{M}_\mathbb{D}$-injective. Using these, we can obtain $h'\colon RC' \to RLD$ such that $h' \circ Rg = Rf$. Since R is full, there is $h\colon C' \to LD$ such that $Rh = h'$. The faithfulness of R implies $h \circ g = f$. Thus LD is $\mathcal{M}_\mathbb{C}$-injective. $\square$

Using this result, we can identify c-injective objects in a few situations.

Example 29 (continuous lattices in **Top** $\to$ **Set** *[30]).* In the setting of Proposition 28, consider the case where $\mathbb{D} =$ **Top**, $\mathbb{C} =$ **Top**$_0$. Here **Top**$_0$ is the full subcategory of **Top** of T_0 spaces. Let R be the inclusion. It has a left adjoint L, taking each space to its Kolmogorov quotient. Let $\mathcal{M}_\mathbb{C}$ be the class of topological embeddings (i.e. homeomorphisms to their images) and $\mathcal{M}_\mathbb{D}$ be the class of Cartesian arrows (w.r.t. the fibration **Top** $\to$ **Set**). Then the assumptions in Proposition 28 are satisfied and we can conclude that c-injective objects in **Top** are precisely injective objects in **Top**$_0$ w.r.t. embeddings.

The latter has been identified by Scott [30]. According to his result, such objects are precisely *continuous lattices* with the Scott topology. Thus, we can see that c-injective objects in **Top** are precisely such spaces.

Example 30 (complete lattices in **Pre** $\to$ **Set** *[4]).* In the setting of Proposition 28, consider the case where $\mathbb{D} =$ **Pre**, $\mathbb{C} =$ **Pos**. Here **Pos** is the full subcategory of **Pre** of posets. Let R be the inclusion. It has a left adjoint L, taking each preordered set to its poset reflection. Let $\mathcal{M}_\mathbb{C}$ be the class of embeddings and $\mathcal{M}_\mathbb{D}$ be the class of Cartesian arrows (w.r.t. the fibration **Pre** $\to$ **Set**). Then the assumptions in Proposition 28 are satisfied and we can conclude that c-injective objects in **Pre** are precisely injective objects in **Pos** w.r.t. embeddings.

The latter has been identified by Banaschewski and Bruns [4]. According to their result, such objects are precisely complete lattices. Thus, we can see that c-injective objects in **Pre** are precisely complete lattices.

4.2   Results Specific to C-Injective Objects

To develop the theory of c-injective objects further, we establish some preservation results for c-injectivity. Based on the two propositions of the last section, we show two propositions specific to fibrations and c-injective objects.

From Proposition 27, we can derive the following:

Proposition 31. Let $p\colon \mathbb{E} \to \mathbb{C}, q\colon \mathbb{F} \to \mathbb{D}$ be $\mathbf{CLat}_\sqcap$-fibrations and $L \dashv R\colon \mathbb{E} \to \mathbb{F}$ be a pair of adjoint functors. If L is fibered (from q to p), then $RE \in \mathbb{F}$ is c-injective (in q) for each c-injective $E \in \mathbb{E}$.

Proof. Let $\mathcal{M}_\mathbb{E}$ be the class of all arrows Cartesian w.r.t. p and $\mathcal{M}_\mathbb{F}$ be the class of all arrows Cartesian w.r.t. q. Then, use Proposition 27 to the pair $L \dashv R$ of adjoint functors.   □

From Proposition 28, we can derive the following:

Proposition 32. In the setting of Proposition 31, assume in addition that both L and R are fibered and that $\eta\colon \mathrm{Id} \to RL$ is componentwise epi. Then, $F \in \mathbb{F}$ is c-injective if and only if it is isomorphic to RE for some c-injective $E \in \mathbb{E}$.

Proof. Use Proposition 28 in the same setting as the proof of Proposition 31. □

5   Examples

We list several examples of Theorem 20. Indeed, most of the examples listed in [18, Table VI] turn out to be fibered by Theorem 20. Since the conditions in Theorem 20 only refer to $p\colon \mathbb{E} \to \mathbb{C}$ and $\mathbf{\Omega}$, we sort the examples by these data.

We here recall some basic functors considered:

Definition 33. Let $\mathcal{P}\colon \mathbf{Set} \to \mathbf{Set}$ be the covariant powerset functor and $\mathcal{D}_{\leq 1}\colon \mathbf{Set} \to \mathbf{Set}$ be the subdistribution functor. Here, a subdistribution $p \in \mathcal{D}_{\leq 1}X$ is a measure on the σ-algebra of all subsets of X with total mass ≤ 1. We abbreviate $p(\{x\})$ to $p(x)$.

5.1   Kantorovich Lifting

In Example 18 we have seen that, in the fibration $\mathbf{PMet}_\top \to \mathbf{Set}$, the object $([0, \top], d_\mathbb{R})$ is c-injective. We gather examples of this case here. As mentioned in Example 22 and Example 25, this class of examples has been already studied and shown to be fibered in [3, 19].

Example 34 (Hausdorff pseudometric). Let $\inf\colon \mathcal{P}[0, \top] \to [0, \top]$ be the map taking any set to its infimum. Then, the codensity lifting $\mathcal{P}^{([0,\top], d_\mathbb{R}), \inf}\colon \mathbf{PMet}_\top \to \mathbf{PMet}_\top$ turns out to induce the *Hausdorff distance*: for any $(X, d_X) \in \mathbf{PMet}_\top$, if we let $(\mathcal{P}X, d_{\mathcal{P}X}) = \mathcal{P}^{([0,\top], d_\mathbb{R}), \inf}(X, d_X)$, then

$$d_{\mathcal{P}X}(S, T) = \max\left(\sup_{x \in S} \inf_{y \in T} d_X(x, y), \sup_{y \in T} \inf_{x \in S} d_X(x, y)\right)$$

holds for any $S, T \in \mathcal{P}X$. By Theorem 20, this functor is fibered.

Example 35 (Kantorovich pseudometric). Let $e \colon \mathcal{D}_{\leq 1}[0, \top] \to [0, \top]$ be the map taking any distribution to its expected value. Then, the codensity lifting

$$\mathcal{D}_{\leq 1}{}^{([0,\top],d_{\mathbb{R}}),e} \colon \mathbf{PMet}_{\top} \to \mathbf{PMet}_{\top}$$

turns out to induce the *Kantorovich distance*: for any $(X, d_X) \in \mathbf{PMet}_{\top}$, if we let $(\mathcal{D}_{\leq 1}X, d_{\mathcal{D}_{\leq 1}X}) = \mathcal{D}_{\leq 1}{}^{([0,\top],d_{\mathbb{R}}),e}(X, d_X)$, then

$$d_{\mathcal{D}_{\leq 1}X}(p, q) = \sup_{f \colon (X,d_X) \to ([0,\top],d_{\mathbb{R}}) \text{ nonexpansive}} \left| \sum_{x \in X} f(x)p(x) - \sum_{x \in X} f(x)q(x) \right|$$

holds for any $p, q \in \mathcal{D}_{\leq 1}X$. By Theorem 20, this functor is fibered.

5.2 Lower, Upper, and Convex Preorders

In Example 30, we have identified complete lattices as c-injective objects in the fibration $\mathbf{Pre} \to \mathbf{Set}$. In particular, the two-point set $(2, \leq)$ is a c-injective object (Example 17).

Katsumata and Sato [16, Sect. 3.1] used codensity lifting to recover the *lower*, *upper*, and *convex preorders* on powersets. Here we see that our result applies to them: all of the following liftings are fibered.

Example 36 (lower preorder). Define $\Diamond \colon \mathcal{P}2 \to 2$ so that $\Diamond S = \top$ if and only if $\top \in S$. Then, the codensity lifting $\mathcal{P}^{(2,\leq),\Diamond} \colon \mathbf{Pre} \to \mathbf{Pre}$ turns out to induce the *lower preorder*: if we let $(\mathcal{P}X, \leq^{\Diamond}_{\mathcal{P}X}) = \mathcal{P}^{(2,\leq),\Diamond}(X, \leq_X)$, then, for any $S, T \in \mathcal{P}X$,

$$S \leq^{\Diamond}_{\mathcal{P}X} T \Leftrightarrow \forall x \in S, \exists y \in T, x \leq_X y.$$

Example 37 (upper preorder). Define $\Box \colon \mathcal{P}2 \to 2$ so that $\Box S = \top$ if and only if $\bot \notin S$. Then, the codensity lifting $\mathcal{P}^{(2,\leq),\Box} \colon \mathbf{Pre} \to \mathbf{Pre}$ turns out to induce the *upper preorder*: if we let $(\mathcal{P}X, \leq^{\Box}_{\mathcal{P}X}) = \mathcal{P}^{(2,\leq),\Box}(X, \leq_X)$, then, for any $S, T \in \mathcal{P}X$,

$$S \leq^{\Box}_{\mathcal{P}X} T \Leftrightarrow \forall y \in T, \exists x \in S, x \leq_X y.$$

Example 38 (convex preorder). Denote the family of the two lifting parameters above by $((2, \leq), \{\Diamond, \Box\})$. Then, the codensity lifting (with multiple parameters, Definition 23) $\mathcal{P}^{(2,\leq),\{\Diamond,\Box\}} \colon \mathbf{Pre} \to \mathbf{Pre}$ is simply the meet of $\mathcal{P}^{(2,\leq),\Diamond}$ and $\mathcal{P}^{(2,\leq),\Box}$. This is what is called the *convex preorder*.

Remark 39. The original formulation [16, Sect. 3.1] is based on codensity lifting of monads, so apparently different to ours. In our terms, they used the multiplication $\mu_1 \colon \mathcal{P}\mathcal{P}1 \to \mathcal{P}1$ and two different preorders on $\mathcal{P}1$. Using two different bijections between $\mathcal{P}1$ and 2, it can be shown that their formulation is actually equivalent to ours.

5.3 Equivalence Relations

In Example 16 we have seen that, in the fibration **EqRel** → **Set**, the object $(2, =)$ is c-injective. We gather examples of this case here. All of the following liftings are fibered. Details on the following examples can be found in [18].

Example 40 (lifting for bisimilarity on Kripke frames). Consider the codensity lifting $\mathcal{P}^{(2,=),\Diamond}$: **EqRel** → **EqRel**, where $\Diamond$ is as defined in Example 36. This turns out to satisfy the following: if we let $(\mathcal{P}X, \sim_{\mathcal{P}X}) = \mathcal{P}^{(2,=),\Diamond}(X, \sim_X)$, then

$$S \sim_{\mathcal{P}X} T \Leftrightarrow (\forall x \in S, \exists y \in T, x \sim_X y) \wedge (\forall y \in T, \exists x \in S, x \sim_X y)$$

holds for any $S, T \in \mathcal{P}X$. This can be used to define (the conventional notion of) bisimilarity on Kripke frames ($\mathcal{P}$-coalgebras).

Example 41 (lifting for bisimilarity on Markov chains). For each $r \in [0, 1]$, define a map $\mathrm{thr}_r \colon \mathcal{D}_{\leq 1} 2 \to 2$ so that $\mathrm{thr}_r(p) = \top$ if and only if $p(\top) \geq r$. These define a $[0, 1]$-indexed family of lifting parameters $((2, =), \mathrm{thr}_r)_{r \in [0,1]}$. The codensity lifting $\mathcal{D}_{\leq 1}^{(2,=),\mathrm{thr}}$ defined by this family can be used to define probabilistic bisimilarity on Markov chains ($\mathcal{D}_{\leq 1}$-coalgebras).

5.4 Topologies

In Example 29, we have identified c-injective objects in the fibration **Top** → **Set**. In particular, the *Sierpinski space*, defined as follows, is a c-injective object:

Definition 42 (Sierpinski space). The *Sierpinski space* is a topological space $(2, \mathcal{O}_\mathbb{O})$ where $2 = \{\bot, \top\}$ and the family $\mathcal{O}_\mathbb{O}$ of open sets is $\{\emptyset, \{\top\}, 2\}$. We denote this space by $\mathbb{O}$.

The following liftings of $\mathcal{P}$ have appeared in [16, Sect. 3.2]. All of them are fibered: in other words, they send embeddings to embeddings.

Example 43 (lower Vietoris lifting). Consider the codensity lifting $\mathcal{P}^{\mathbb{O},\Diamond}$: **Top** → **Top**, where $\Diamond$ is as defined in Example 36. For each $(X, \mathcal{O}_X) \in$ **Top**, if we let $(\mathcal{P}X, \mathcal{O}_{\mathcal{P}X}^\Diamond) = \mathcal{P}^{\mathbb{O},\Diamond}(X, \mathcal{O}_X)$, then the topology $\mathcal{O}_{\mathcal{P}X}^\Diamond$ is the coarsest one such that, for each $U \in \mathcal{O}_X$, the set $\{V \subseteq X \mid V \cap U \neq \emptyset\}$ is open. This is called *lower Vietoris lifting* in [16].

Example 44 (upper Vietoris lifting). Consider the codensity lifting $\mathcal{P}^{\mathbb{O},\Box}$: **Top** → **Top**, where $\Box$ is as defined in Example 37. For each $(X, \mathcal{O}_X) \in$ **Top**, if we let $(\mathcal{P}X, \mathcal{O}_{\mathcal{P}X}^\Box) = \mathcal{P}^{\mathbb{O},\Box}(X, \mathcal{O}_X)$, then the topology $\mathcal{O}_{\mathcal{P}X}^\Box$ is the coarsest one such that, for each $U \in \mathcal{O}_X$, the set $\{V \subseteq X \mid V \subseteq U\}$ is open. This is called *upper Vietoris lifting* in [16].

Example 45 (Vietoris lifting). Define the codensity lifting $\mathcal{P}^{\mathbb{O},\{\Diamond,\Box\}}$: **Top** → **Top** like one in Example 38. We call this *Vietoris lifting*.

This turns out to be connected to *Vietoris topology* [20] as follows. For each $(X, \mathcal{O}_X) \in$ **Top**, let $(\mathcal{P}X, \mathcal{O}_{\mathcal{P}X}^{\Diamond,\Box}) = \mathcal{P}^{\mathbb{O},\{\Diamond,\Box\}}(X, \mathcal{O}_X)$. The set $K(X, \mathcal{O}_X)$ of closed subsets of $(X, \mathcal{O}_X)$ is a subset of $\mathcal{P}X$. Here, the topology on $K(X, \mathcal{O}_X)$ induced from $\mathcal{O}_{\mathcal{P}X}^{\Diamond, \Box}$ is the same as the Vietoris topology.

This coincidence and the fiberedness of $\mathcal{P}^{\mathbb{O}, \{\Diamond, \Box\}}$ implies that the *Vietoris functor* $\mathbb{V}\colon$ **Stone** $\to$ **Stone**, defined in [20], sends embeddings to embeddings.

In [18], we considered another lifting:

Example 46 (lifting for bisimulation topology). Fix any set Σ. Let $A_\Sigma\colon$ **Set** $\to$ **Set** be the functor defined by $A_\Sigma X = 2 \times X^\Sigma$. Define acc$\colon A_\Sigma 2 \to 2$ by $\mathrm{acc}(t, \rho) = t$. For each $a \in \Sigma$, define $\langle a \rangle \colon A_\Sigma 2 \to 2$ by $\langle a \rangle (t, \rho) = \rho(a)$. Here, $(\mathbb{O}, \mathrm{acc})$ and $(\mathbb{O}, \langle a \rangle)$ for each $a \in \Sigma$ consist of a family of lifting parameters. The codensity lifting (with multiple parameters, Definition 23) $A_\Sigma^{\mathbb{O}, \{\mathrm{acc}\} \cup \{\langle a \rangle | a \in \Sigma\}}\colon$ **Top** $\to$ **Top** was used to define *bisimulation topology* for deterministic automata (A_Σ-coalgebras). This is fibered. This fact is used in Example 51, where we will look at bisimulation topology again.

6   Application to Codensity Bisimilarity

Now we present an application of our main result. Based on codensity lifting, we defined *codensity bisimilarity* in [18]. It subsumes bisimilarity, simulation preorder, and behavioral metric as special cases. Here we see that, in the cases to which our fiberedness result applies, codensity bisimilarity interacts well with coalgebra morphisms. In particular, the codensity bisimilarity on any coalgebra is determined by that on the final coalgebra, if it exists.

Recall the definition of coalgebra:

Definition 47 (coalgebra of an endofunctor). Let $F\colon \mathbb{C} \to \mathbb{C}$ be an endofunctor on a category $\mathbb{C}$. An F-*coalgebra* is a pair of an object $X \in \mathbb{C}$ and an arrow $c\colon X \to FX$.

Let $c\colon X \to FX$ and $d\colon Y \to FY$ be F-coalgebras. A *morphism of coalgebras* from (X, c) to (Y, d) is an arrow $f\colon X \to Y$ in $\mathbb{C}$ such that $d \circ f = Ff \circ c$ holds.

As sketched in Sect. 1, functor lifting can be used to define a "bisimilarity-like notion". If we use codensity lifting in this construction, we obtain the following definition:

Definition 48 (codensity bisimilarity [18, Definitions III.6 and III.8]). Assume the setting of Definition 23. Let $c\colon X \to FX$ be any F-coalgebra. Define $\Phi_c^{\Omega, \tau}\colon \mathbb{E}_X \to \mathbb{E}_X$ by $\Phi_c^{\Omega, \tau} P = c^* \left(F^{\Omega, \tau} P \right)$.

The $((\Omega, \tau)\text{-})$*codensity bisimilarity* is the greatest fixed point (w.r.t. $\sqsubseteq$) of $\Phi_c^{\Omega, \tau}$. We denote this by $\nu \Phi_c^{\Omega, \tau}$.

Note that the greatest fixed point of $\Phi_c^{\Omega, \tau}$ always exists. This can be seen, for example, by the Tarski fixed point theorem. Another option is to use the constructive fixed point theorem by Cousot and Cousot [6]. We use their characterization of the greatest fixed point to prove the following proposition:

Proposition 49 (stability of codensity bisimilarity). Assume the setting of Definition 23 (codensity lifting with multiple parameters). Assume also that each Ω_a is a c-injective object. Then, codensity bisimilarity is stable under coalgebra morphisms: for any morphism of coalgebras f from (X, c) to (Y, d), we have $\nu\Phi_c^{\Omega,\tau} = f^*\left(\nu\Phi_d^{\Omega,\tau}\right)$.

Proof. Define a transfinite sequence $\left(\nu_\alpha\Phi_c^{\Omega,\tau}\right)_{\alpha \text{ is an ordinal}}$ of elements of $\mathbb{E}_X$ by the following:

$$\nu_\alpha\Phi_c^{\Omega,\tau} = \prod_{\beta < \alpha} \Phi_c^{\Omega,\tau}\left(\nu_\beta\Phi_c^{\Omega,\tau}\right).$$

Define another transfinite sequence $\left(\nu_\alpha\Phi_d^{\Omega,\tau}\right)_{\alpha \text{ is an ordinal}}$ by a similar manner. By the result in [6], there is an ordinal γ such that $\nu_\gamma\Phi_c^{\Omega,\tau} = \nu\Phi_c^{\Omega,\tau}$ and $\nu_\gamma\Phi_d^{\Omega,\tau} = \nu\Phi_d^{\Omega,\tau}$.[1] Thus, it suffices to show the following claim:

Claim. For any ordinal α, we have $\nu_\alpha\Phi_c^{\Omega,\tau} = f^*\left(\nu_\alpha\Phi_d^{\Omega,\tau}\right)$.

We show this by transfinite induction on α. Assume the claim holds for all $\beta < \alpha$.

Using the assumption that f is a morphism of coalgebras, the fiberedness of $F^{\Omega,\tau}$ (Corollary 24), and the functoriality of pullback (Proposition 5), we have $f^* \circ \Phi_d^{\Omega,\tau} = \Phi_c^{\Omega,\tau} \circ f^*$. It implies the claim for α

$$f^*\left(\nu_\alpha\Phi_d^{\Omega,\tau}\right) = f^*\left(\prod_{\beta<\alpha}\Phi_d^{\Omega,\tau}\left(\nu_\beta\Phi_d^{\Omega,\tau}\right)\right) = \prod_{\beta<\alpha}f^*\left(\Phi_d^{\Omega,\tau}\left(\nu_\beta\Phi_d^{\Omega,\tau}\right)\right)$$

$$= \prod_{\beta<\alpha}\Phi_c^{\Omega,\tau}\left(f^*\nu_\beta\Phi_d^{\Omega,\tau}\right) = \prod_{\beta<\alpha}\Phi_c^{\Omega,\tau}\left(\nu_\beta\Phi_c^{\Omega,\tau}\right)$$

$$= \nu_\alpha\Phi_c^{\Omega,\tau}.$$

$\square$

In particular, the codensity bisimilarity is determined by that on the final coalgebra:

Corollary 50. Assume the setting of Proposition 49. Assume also that there exists a final F-coalgebra $z\colon Z \to FZ$. Then, for any F-coalgebra $c\colon X \to FX$, the unique coalgebra morphism $!_X\colon X \to Z$ satisfies $\nu\Phi_c^{\Omega,\tau} = (!_X)^*\left(\nu\Phi_z^{\Omega,\tau}\right)$.

Example 51 (bisimulation topology for deterministic automata). Recall Example 46. For any A_Σ-coalgebra $c\colon X \to A_\Sigma X$, we defined the codensity bisimilarity on X by $\nu\Phi_c^{\mathbb{0},\{\text{acc}\}\cup\{\langle a\rangle|a\in\Sigma\}} \in \mathbf{Top}_X$ [18].

---

[1] This formulation differs slightly from the conventional one where successor and limit ordinals are distinguished, but the result also holds under this definition.

The functor A_Σ has a final coalgebra: the set 2^{Σ^*} of all languages on the alphabet Σ can be given an A_Σ-coalgebra structure and it is final. For an A_Σ-coalgebra $c \colon X \to A_\Sigma X$, the unique coalgebra morphism $l \colon X \to 2^{\Sigma^*}$ assigns to each state the recognized language when started from it.

Corollary 50 implies that this map l determines the bisimulation topology on X. We believe that this fact is new, and it supports our use of the term *language topology* in [18, §VIII-C].

7   Conclusions

Inspired by the proof of fiberedness of Kantorovich lifting [3], we showed a sufficient condition for codensity lifting to be fibered. We listed a number of examples that satisfy the sufficient condition. In addition, we apply the fiberedness to show a result on codensity bisimilarity.

One possible direction of research is to investigate the notion of c-injectiveness in more depth. The existing work on injective objects in homological algebra and topos theory can be a clue for that. In particular, we have not studied which category has *enough c-injectives*. This may be connected with some deep fibrational property.

Another possible direction is to generalize the main result. In [16], codensity lifting of a monad was introduced for a general fibration in terms of right Kan extension. This definition can readily be adapted to endofunctors, but in the current paper, we considered only $\mathbf{CLat}_\sqcap$-fibrations. Extending the main result to this general situation, in particular, to non-poset fibrations, may broaden the scope of application. It can also be fruitful to extend the definition of codensity lifting itself: for example, in Definition 13, we could substitute $\tau^*\Omega$ with other objects above $F\Omega$.[2] Seeking consequences and examples of this version of the definition is future work. Another related research direction is to obtain a similar sufficient condition for fiberedness of *categorical $\top\top$-lifting* [15].

Last but not least, we have to seek other applications. As mentioned in Sect. 1, functor lifting is used in many situations. Using codensity lifting there and seeing what can be implied by the current result seems to be a promising research direction. In particular, codensity lifting seems to be intimately connected to *coalgebraic modal logic*, where $\tau \colon F\Omega \to \Omega$ is regarded as a *modality*. Recently, Kupke and Rot [21] have identified a sufficient condition for a logic to expressive w.r.t. a coinductive predicate (like bisimilarity, behavioral metric etc.). They used fiberedness of lifting in a crucial way (they use the term *fibration map*), which suggests that the current work can play a pivotal role in investigating modal logics.

Acknowledgments. The author is grateful to Ichiro Hasuo and Shin-ya Katsumata for fruitful discussions on technical and structural points. The author is also indebted to anonymous reviewers for clarifying things and pointing out possible future directions, including a topos-theoretic viewpoint and an alternative definition of codensity lifting.

---

[2] This has been pointed out by an anonymous reviewer.

References

1. Baer, R.: Abelian groups that are direct summands of every containing abelian group. Bull. Am. Math. Soc. **46**(10), 800–806 (1940). https://projecteuclid.org: 443/euclid.bams/1183503234
2. Baldan, P., Bonchi, F., Kerstan, H., König, B.: Behavioral Metrics via Functor Lifting. In: Raman, V., Suresh, S.P. (eds.) 34th International Conference on Foundation of Software Technology and Theoretical Computer Science (FSTTCS 2014). Leibniz International Proceedings in Informatics (LIPIcs), vol. 29, pp. 403–415. Schloss Dagstuhl-Leibniz-Zentrum fuer Informatik, Dagstuhl, Germany (2014). https://doi.org/10.4230/LIPIcs.FSTTCS.2014.403, http://drops.dagstuhl. de/opus/volltexte/2014/4859
3. Baldan, P., Bonchi, F., Kerstan, H., König, B.: Coalgebraic behavioral metrics. Log. Methods Comput. Sci. **14**(3) (2018). https://doi.org/10.23638/LMCS-14(3: 20)2018
4. Banaschewski, B., Bruns, G.: Categorical characterization of the macneille completion. Archiv der Mathematik **18**(4), 369–377 (1967). https://doi.org/10.1007/ BF01898828
5. Chatzikokolakis, K., Gebler, D., Palamidessi, C., Xu, L.: Generalized bisimulation metrics. In: Baldan, P., Gorla, D. (eds.) CONCUR 2014. LNCS, vol. 8704, pp. 32–46. Springer, Heidelberg (2014). https://doi.org/10.1007/978-3-662-44584-6_4
6. Cousot, P., Cousot, R.: Constructive versions of tarski's fixed point theorems. Pacific J. Math. **82**(1), 43–57 (1979), https://projecteuclid.org:443/euclid.pjm/ 1102785059
7. Espínola, R., Khamsi, M.A.: Introduction to hyperconvex spaces. In: Kirk, W.A., Sims, B. (eds.) Handbook of Metric Fixed Point Theory, pp. 391–435. Springer, Netherlands (2001). https://doi.org/10.1007/978-94-017-1748-9_13
8. Fujii, S.: Enriched categories and tropical mathematics (2019)
9. Hasuo, I., Cho, K., Kataoka, T., Jacobs, B.: Coinductive predicates and final sequences in a fibration. Electron. Notes Theor. Comput. Sci. **298**, 197–214 (2013). https://doi.org/10.1016/j.entcs.2013.09.014
10. Hermida, C.: Fibrations, logical predicates and indeterminates. Ph.D. thesis, University of Edinburgh, UK (1993)
11. Hermida, C., Jacobs, B.: Structural induction and coinduction in a fibrational setting. Information and Computation **145**(2), 107–152 (1998). https://doi. org/10.1006/inco.1998.2725, http://www.sciencedirect.com/science/article/pii/ S0890540198927250
12. Hilton, P.J., Stammbach, U.: A Course in Homological Algebra, Graduate Texts in Mathematics, vol. 4. Springer, New York (1997). https://doi.org/10.1007/978- 1-4419-8566-8
13. Jacobs, B.: Categorical Logic and Type Theory. No. 141 in Studies in Logic and the Foundations of Mathematics, North Holland, Amsterdam (1999)
14. Kashiwara, M., Schapira, P.: Categories and Sheaves, Grundlehren der mathematischen Wissenschaften, vol. 332. Springer-Verlag, Berlin Heidelberg (2006). https:// doi.org/10.1007/3-540-27950-4
15. Katsumata, S.: A Semantic Formulation of tt-Lifting and Logical Predicates for Computational Metalanguage. In: Ong, L. (ed.) CSL 2005. LNCS, vol. 3634, pp. 87–102. Springer, Heidelberg (2005). https://doi.org/10.1007/11538363_8

16. Katsumata, S., Sato, T.: Codensity liftings of monads. In: Moss, L.S., Sobocinski, P. (eds.) 6th Conference on Algebra and Coalgebra in Computer Science (CALCO 2015). Leibniz International Proceedings in Informatics (LIPIcs), vol. 35, pp. 156–170. Schloss Dagstuhl-Leibniz-Zentrum fuer Informatik, Dagstuhl, Germany (2015). https://doi.org/10.4230/LIPIcs.CALCO.2015.156, http://drops. dagstuhl.de/opus/volltexte/2015/5532

17. Klin, B.: The least fibred lifting and the expressivity of coalgebraic modal logic. In: Fiadeiro, J.L., Harman, N., Roggenbach, M., Rutten, J. (eds.) CALCO 2005. LNCS, vol. 3629, pp. 247–262. Springer, Heidelberg (2005). https://doi.org/10. 1007/11548133_16

18. Komorida, Y., Katsumata, S., Hu, N., Klin, B., Hasuo, I.: Codensity games for bisimilarity. In: 2019 34th Annual ACM/IEEE Symposium on Logic in Computer Science (LICS), June 2019. https://doi.org/10.1109/LICS.2019.8785691

19. König, B., Mika-Michalski, C.: (Metric) Bisimulation Games and Real-Valued Modal Logics for Coalgebras. In: Schewe, S., Zhang, L. (eds.) 29th International Conference on Concurrency Theory (CONCUR 2018). Leibniz International Proceedings in Informatics (LIPIcs), vol. 118, pp. 37:1–37:17. Schloss Dagstuhl-Leibniz-Zentrum fuer Informatik, Dagstuhl, Germany (2018). https://doi. org/10.4230/LIPIcs.CONCUR.2018.37, http://drops.dagstuhl.de/opus/volltexte/ 2018/9575

20. Kupke, C., Kurz, A., Venema, Y.: Stone coalgebras. Electr. Notes Theor. Comput. Sci. **82**(1), 170–190 (2003). https://doi.org/10.1016/S1571-0661(04)80638-8

21. Kupke, C., Rot, J.: Expressive logics for coinductive predicates, to appear in Proceedings CSL2020 (2020)

22. Leinster, T.: Codensity and the ultrafilter monad. TAC (2013)

23. Leinster, T.: Basic Category Theory. Cambridge Studies in Advanced Mathematics, Cambridge University Press (2014). https://doi.org/10.1017/CBO9781107360068

24. Mac Lane, S.: Abelian categories. Categories for the Working Mathematician. GTM, vol. 5, pp. 191–209. Springer, New York (1978). https://doi.org/10.1007/ 978-1-4757-4721-8_9

25. Milner, R.: Communication and Concurrency. Prentice-Hall, Upper Saddle River (1989)

26. Park, D.: Concurrency and automata on infinite sequences. In: Deussen, P. (eds.) Proceedings of the 5th GI-Conference on Theoretical Computer Science, vol 104 pp. 167–183. Springer, London (1981). https://doi.org/10.1007/BFb0017309, http:// dl.acm.org/citation.cfm?id=647210.720030

27. Rutten, J.J.M.M.: Universal coalgebra: a theory of systems. Theor. Comput. Sci. **249**(1), 3–80 (2000). https://doi.org/10.1016/S0304-3975(00)00056-6

28. Sangiorgi, D.: Introduction to Bisimulation and Coinduction. Cambridge University Press, Cambridge (2011). https://doi.org/10.1017/CBO9780511777110

29. Sato, T., Barthe, G., Gaboardi, M., Hsu, J., Katsumata, S.: Approximate span liftings: Compositional semantics for relaxations of differential privacy. In: 34th Annual ACM/IEEE Symposium on Logic in Computer Science, LICS 2019, Vancouver, BC, Canada, 24–27 June 2019, pp. 1–14. IEEE (2019). https://doi.org/10. 1109/LICS.2019.8785668

30. Scott, D.: Continuous lattices. In: Lawvere, F.W. (ed.) Toposes, Algebraic Geometry and Logic. LNM, vol. 274, pp. 97–136. Springer, Heidelberg (1972). https:// doi.org/10.1007/BFb0073967

31. Sprunger, D., Katsumata, S., Dubut, J., Hasuo, I.: Fibrational bisimulations and quantitative reasoning. In: Cîrstea, C. (ed.) CMCS 2018. LNCS, vol. 11202, pp. 190–213. Springer, Cham (2018). https://doi.org/10.1007/978-3-030-00389-0_11

Explaining Non-bisimilarity
in a Coalgebraic Approach: Games
and Distinguishing Formulas

Barbara König[1] , Christina Mika-Michalski[1(✉)], and Lutz Schröder[2]

[1] University of Duisburg-Essen, Duisburg, Germany
{barbara_koenig,christina.mika-michalski}@uni-due.de
[2] Friedrich-Alexander-Universität Erlangen-Nürnberg, Erlangen, Germany
lutz.schroeder@fau.de

Abstract. Behavioural equivalences can be characterized via bisimulation, modal logics, and spoiler-duplicator games. In this paper we work in the general setting of coalgebra and focus on generic algorithms for computing the winning strategies of both players in a bisimulation game. The winning strategy of the spoiler (if it exists) is then transformed into a modal formula that distinguishes the given non-bisimilar states. The modalities required for the formula are also synthesized on-the-fly, and we present a recipe for re-coding the formula with different modalities, given by a separating set of predicate liftings. Both the game and the generation of the distinguishing formulas have been implemented in a tool called T-BEG.

Keywords: Coalgebra · Bisimulation games · Distinguishing formulas · Generic partition refinement

1 Introduction

There are many contexts in which it is useful to check whether two system states are behaviourally equivalent respectively bisimilar. In this way one can compare a system with its specification, replace a subsystem by another one that is behaviourally equivalent or minimize a transition system. Here we will concentrate on methods for explaining that two given states in a transition system are *not* bisimilar. The idea is to provide a witness for non-bisimilarity. Such a witness can be used to explain (to the user) why an implementation does not conform to a specification and give further insights for adjusting it.

Two states are bisimilar if they are related by a bisimulation relation. But this definition does not provide us with an immediate witness for non-bisimilarity, since we would have to enumerate all relations including that particular pair of states and show that they are not bisimulations. Hence, we have to resort

Work by the first two authors supported by the DFG project BEMEGA (KO 2185/7-2). Work by the third author forms part of the DFG project ProbDL2 (SCHR 1118/6-2).

ⓒ IFIP International Federation for Information Processing 2020
Published by Springer Nature Switzerland AG 2020
D. Petrişan and J. Rot (Eds.): CMCS 2020, LNCS 12094, pp. 133–154, 2020.
https://doi.org/10.1007/978-3-030-57201-3_8

to other characterizations of bisimilarity: bisimulation games [27], also known as spoiler-duplicator games, and modal logic. In the former case a proof of the non-bisimilarity of two states is given by a winning strategy of the spoiler. In the latter case the Hennessy-Milner theorem [14] guarantees for image-finite labelled transition systems that, given two non-bisimilar states x_0, x_1, there exists a modal formula φ such that one of the states satisfies φ and the other does not. The computation of such distinguishing formulas is explained in [6].

While the results and techniques above have been introduced for labelled transition systems, we are here interested in the more general setting of coalgebras [25], which encompass various types of transition systems. Here we concentrate on coalgebras living in **Set**, where an endofunctor $F \colon \mathbf{Set} \to \mathbf{Set}$ specifies the branching type of the coalgebra (non-deterministic, probabilistic, etc.).

Modal logics have been extensively studied for coalgebras and it has been shown that under certain restrictions, modal coalgebraic logic is expressive, i.e., it satisfies the Hennessy-Milner theorem [23,26]. However, to our knowledge, no explicit construction of distinguishing formulas in the coalgebraic setting has yet been given.

Coalgebraic games have been studied to a lesser extent: we refer to Baltag [2,3], where the game is based on providing subsets of bisimulation relations (under the assumption that the functor F is weak pullback preserving) and a generalization of Baltag's game to other functors in [21]. Furthermore there is our own contribution [18], on which this article is based, and [16], which considers codensity games from an abstract, fibrational perspective.

We combine both the game and the modal logic view on coalgebras and present the following contributions:

▷ *We describe how to compute the winning strategies of the players in the behavioural equivalence game.*

▷ *We show how to construct a distinguishing formula based on the spoiler strategy. The modalities for the formula are not provided a priori, but are synthesized on-the-fly as so-called* cone modalities *while generating the formula.*

▷ *Finally we show under which conditions one can re-code a formula with such modalities into a formula with different modalities, given by a separating set of predicate liftings.*

Both the game and the generation of the distinguishing formulas have been implemented in a generic tool called T-BEG[1], where the functor is provided as a parameter. In particular, using this tool, one can visualize coalgebras, play the game (against the computer), derive winning strategies and convert the winning strategy of the spoiler into a distinguishing formula. Since the development of the tool was our central aim, we have made design decisions in such a way that we obtain effective algorithms. This means that we have taken a hands-on approach and avoided constructions that potentially iterate over infinitely many elements (such as the set of all modalities, which might be infinite). The partition refinement algorithm presented in the paper distinguishes states that are not behaviourally equivalent by a single equivalence class compared to other

---

[1] Available at: https://www.uni-due.de/theoinf/research/tools_tbeg.php.

techniques which iterate over the final chain [9,17]. Separation via a single equivalence class is a technique used within known algorithms for checking bisimilarity in labelled transition systems [15,22]. This requires a certain assumption on the endofunctor specifying the branching type (dubbed *separability by singletons*). Note that [22] has already been generalized to a coalgebraic setting in [9], using the assumption of *zippability*. Here we compare these two assumptions.

After presenting the preliminaries (Sect. 2), including the game, we describe how to compute the winning strategies in Sect. 3. In Sect. 4 we show how to construct and re-code distinguishing formulas, followed by a presentation of the tool T-BEG in Sect. 5. Finally, we conclude in Sect. 6. The proofs can be found in the full version of the paper [19].

2   Preliminaries

Equivalence Relations and Characteristic Functions: Let $R \subseteq X \times X$ be an *equivalence relation*, where the set of all equivalence relations on X is given by $Eq(X)$. For $x_0 \in X$ we denote the *equivalence class* of x_0 by $[x_0]_R = \{x_1 \in X \mid (x_0, x_1) \in R\}$. By $E(R)$ we denote the set of all equivalence classes of R. Given $Y \subseteq X$, we define the *R-closure* of Y as follows: $[Y]_R = \{x_1 \in X \mid \exists x_0 \in Y (x_0, x_1) \in R\}$.

For $Y \subseteq X$, we denote its *predicate* or *characteristic function* by $\chi_Y \colon X \to \{0, 1\}$. Furthermore, given a characteristic function $\chi \colon X \to \{0, 1\}$, its corresponding set is denoted $\hat{\chi} \subseteq X$.

We will sometimes overload the notation and for instance write $[p]_R$ for the R-closure of a predicate p. Furthermore we will write $p_0 \cap p_1$ for the intersection of two predicates.

Coalgebra: We restrict our setting to the category **Set**, in particular we assume an *endofunctor* $F \colon$ **Set** $\to$ **Set**, intuitively describing the branching type of the transition system under consideration. A *coalgebra* [25], describing a transition system of this branching type, is given by a function $\alpha \colon X \to FX$. Two states $x_0, x_1 \in X$ are *behaviourally equivalent* ($x_0 \sim x_1$) if there exists a coalgebra homomorphism f from α to some coalgebra $\beta \colon Y \to FY$ (i.e., a function $f \colon X \to Y$ with $\beta \circ f = Ff \circ \alpha$) such that $f(x_0) = f(x_1)$. We assume that F preserves weak pullbacks, which means that behavioural equivalence and coalgebraic bisimilarity coincide, and we will use the two terms interchangeably.

Preorder Lifting: Furthermore we need to *lift* preorders under a functor F. To this end, we use the lifting introduced in [1] (essentially the standard *Barr extension* of F [4,28]), which guarantees that the lifted relation is again a preorder provided that F preserves weak pullbacks: Let $\leq$ be a preorder on Y, i.e. $\leq \subseteq Y \times Y$. We define a preorder $\leq^F$ on FY by $t_0 \leq^F t_1$ iff there exists $t \in F(\leq)$ such that $F\pi_i(t) = t_i$ for $i \in \{0, 1\}$, where $\pi_i \colon \leq \to Y$ are the usual projections. More concretely, we consider the order $\leq = \{(0, 0), (0, 1), (1, 1)\}$ over $2 = \{0, 1\}$ and its corresponding liftings $\leq^F$.

Note that applying the functor is monotone wrt. the lifted order:

Lemma 1 ([18]). *Let $(Y, \leq)$ be an ordered set and let $p_0, p_1 \colon X \to Y$ be functions. Then $p_0 \leq p_1$ implies $Fp_0 \leq^F Fp_1$, with both inequalities read pointwise.*

Predicate Liftings: In order to define the modal logic, we need the notion of *predicate liftings* (also called *modalities*). Formally, a predicate lifting for F is a natural transformation $\bar{\lambda} \colon \mathcal{Q} \Rightarrow \mathcal{Q}F$, where $\mathcal{Q}$ is the contravariant powerset functor. It transforms subsets $P \subseteq X$ into subsets $\bar{\lambda}(P) \subseteq FX$.

We use the fact that predicate liftings are in one-to-one correspondence with functions of type $\lambda \colon F2 \to 2$ (which specify subsets of $F2$ and will also be called *evaluation maps*) [26]. We view subsets $P \subseteq X$ as predicates $p = \chi_P$ and lift them via $p \mapsto \lambda \circ Fp$. In order to obtain expressive logics, we also need the notion of a separating set of predicate liftings.

Definition 2. A set Λ of evaluation maps for a functor $F \colon \mathbf{Set} \to \mathbf{Set}$ is *separating* if for all sets X and $t_0, t_1 \in FX$ with $t_0 \neq t_1$, there exists $\lambda \in \Lambda$ and $p \colon X \to 2$ such that $\lambda(Fp(t_0)) \neq \lambda(Fp(t_1))$.

This means that every $t \in FX$ is uniquely determined by the set $\{(\lambda, p) \mid \lambda \in \Lambda, p \colon X \to 2, \lambda(Fp(t)) = 1\}$. Such a separating set of predicate liftings exists iff $(Fp \colon FX \to F2)_{p \colon X \to 2}$ is jointly injective.

Here we concentrate on *unary* predicate liftings: If one generalizes to *polyadic* predicate liftings, a separating set of predicate liftings can be found for every accessible functor [26].

Separating sets of monotone predicate liftings and the lifted order on $F2$ are related as follows:

Proposition 3 ([18]). *An evaluation map $\lambda \colon F2 \to 2$ corresponds to a monotone predicate lifting $(p \colon X \to 2) \mapsto (\lambda \circ Fp \colon FX \to 2)$ iff $\lambda \colon (F2, \leq^F) \to (2, \leq)$ is monotone.*

Proposition 4 ([18]). *F has a separating set of monotone predicate liftings iff $\leq^F \subseteq F2 \times F2$ is anti-symmetric and $(Fp \colon FX \to F2)_{p \colon X \to 2}$ is jointly injective.*

Coalgebraic Modal Logics: Given a cardinal κ and a set Λ of evaluation maps $\lambda \colon F2 \to 2$, we define a coalgebraic modal language $\mathcal{L}^\kappa(\Lambda)$ via the grammar

$$\varphi ::= \bigwedge \Phi \mid \neg\varphi \mid [\lambda]\varphi \quad \text{where } \Phi \subseteq \mathcal{L}^\kappa(\Lambda) \text{ with } card(\Phi) < \kappa \text{ and } \lambda \in \Lambda.$$

The last case describes the prefixing of a formula φ with a modality $[\lambda]$. Given a coalgebra $\alpha \colon X \to FX$ and a formula φ, the semantics of such a formula is given by a map $[\![\varphi]\!]_\alpha \colon X \to 2$, where conjunction and negation are interpreted as usual and $[\![[\lambda]\varphi]\!]_\alpha = \lambda \circ F[\![\varphi]\!]_\alpha \circ \alpha$.

For simplicity we will often write $[\![\varphi]\!]$ instead of $[\![\varphi]\!]_\alpha$. Furthermore for $x \in X$, we write $x \models \varphi$ whenever $[\![\varphi]\!](x) = 1$. As usual, whenever $[\![\varphi]\!]_\alpha = [\![\psi]\!]_\alpha$ for all coalgebras α we write $\varphi \equiv \psi$. We will use derived operators such as tt (empty conjunction), ff ($\neg tt$) and $\bigvee$ (disjunction).

The logic is always adequate, i.e., two behaviourally equivalent states satisfy the same formulas. Furthermore whenever F is κ-accessible and the set Λ of predicate liftings is separating, it can be shown that the logic is also expressive, i.e., two states that satisfy the same formulas are behaviourally equivalent [24, 26].

Bisimulation Game: We will present the game rules first introduced in [18]. At the beginning of a game, two states x_0, x_1 are given. The aim of the spoiler (S) is to prove that $x_0 \not\sim x_1$, the duplicator (D) attempts to show $x_0 \sim x_1$.

- **Initial configuration:** A coalgebra $\alpha\colon X \to FX$ and a position given as pair $(x_0, x_1) \in X \times X$. From a position (x_0, x_1), the game play proceeds as follows:
- **Step 1:** S chooses $j \in \{0, 1\}$, (i.e. x_0 or x_1) , and a predicate $p_j\colon X \to 2$.
- **Step 2:** D must respond for x_{1-j} with a predicate p_{1-j} satisfying

$$Fp_j(\alpha(x_j)) \leq^F Fp_{1-j}(\alpha(x_{1-j})).$$

- **Step 3:** S chooses $\ell \in \{0, 1\}$ (i.e. p_0 or p_1) and an $x'_\ell \in X$ with $p_\ell(x'_\ell) = 1$.
- **Step 4:** D must respond with an $x'_{1-\ell} \in X$ such that $p_{1-\ell}(x'_{1-\ell}) = 1$.

After one round the game continues in Step 1 with the pair (x'_0, x'_1). D wins if the game continues forever or if S has no move at Step 3. In all other cases, i.e. D has no move at Step 2 or Step 4, S wins.

This game generalizes a bisimulation game for probabilistic transition systems from [8]. Note that – different from the presentation in [8] – we could also restrict the game in such a way that S has to choose index $\ell = 1 - j$ in Step 3.

We now give an example that illustrates the differences between our generic game and the classical bisimulation game for labelled transition systems [27].

Example 5. Consider the transition system in Fig. 1, which depicts a coalgebra $\alpha\colon X \to FX$, where $F = \mathcal{P}_f(A \times (-))$ specifies finitely branching labelled transition systems. Clearly $x_0 \not\sim x_1$.

First consider the classical game where one possible winning strategy of the spoiler is as follows: he moves $x_0 = 1 \xrightarrow{a} 4$, which must be answered by the duplicator via $x_1 = 2 \xrightarrow{a} 5$. Now the spoiler switches and makes a move $5 \xrightarrow{a} 8$, which can not be answered by the duplicator.

In our case a corresponding game proceeds as follows: the spoiler chooses $j = 0$ and $p_0 = \chi_{\{4\}}$. Now the duplicator takes x_1 and can for instance answer with $p_1 = \chi_{\{5\}}$, which leads to

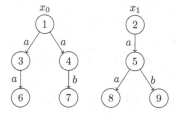

Fig. 1. Spoiler has a winning strategy at (x_0, x_1).

$$Fp_0(\alpha(x_0)) = \{(a, 0), (a, 1)\} \leq^F \{(a, 1)\} = Fp_1(\alpha(x_1))$$

(Compare this with the visualization of the order $\leq^F$ on $F2$ in Fig. 2.) Regardless of how S and D choose states, the next game configuration is $(4, 5)$.

Now the spoiler is *not* forced to switch, but can choose $j = 0$ (i.e. 4) and can play basically any predicate p_0, which leads to either $Fp_0(\alpha(4)) = \{(b, 1)\}$ or $Fp_0(\alpha(4)) = \{(b, 0)\}$. D has no answering move, since $Fp_1(\alpha(5))$ will always contain tuples with a *and* b, which are not in $\leq^F$-relation with the move of S (see also Fig. 2, which depicts $F2$ and its order).

The game characterizes bisimulation for functors that are weak pullback preserving and for which the lifted order $\leq^F$ is anti-symmetric. Then it holds that $x_0 \sim x_1$ if and only if D has a winning strategy from the initial configuration (x_0, x_1). As already shown in [18], in the case of two non-bisimilar states $x_0 \nsim x_1$ we can convert a modal formula φ distinguishing x_0, x_1, i.e., $x_0 \models \varphi$ and $x_1 \not\models \varphi$, into a winning strategy for the spoiler. Furthermore we can extract the winning strategy for the duplicator from the bisimulation relation.

However, in [18] we did not yet show how to directly derive the winning strategy of both players or how to construct a distinguishing formula φ.

3   Computation of Winning Strategies

In the rest of the paper we will fix a coalgebra $\alpha : X \to FX$ with finite X for a weak pullback preserving endofunctor $F \colon \mathbf{Set} \to \mathbf{Set}$. Furthermore we assume that F has a separating set of *monotone* predicate liftings, which implies that $\leq^F$, the lifted order on 2, is anti-symmetric, hence a partial order.

We first present a simple but generic partition refinement algorithm to derive the winning strategy for the spoiler (S) and duplicator (D) for a given coalgebra $\alpha \colon X \to FX$. This is based on a fixpoint iteration that determines those pairs of states $(x_0, x_1) \in X \times X$ for which D has a winning strategy, i.e. $x_0 \sim x_1$. In particular we consider the relation W_α, which – as we will show—is the greatest fixpoint of the following monotone function $\mathcal{F}_\alpha \colon Eq(X) \to Eq(X)$ on equivalence relations:

$$\mathcal{F}_\alpha(R) = \{(x_0, x_1) \in R \mid \forall P \in E(R) \colon F\chi_P(\alpha(x_0)) = F\chi_P(\alpha(x_1))\}$$

$$W_\alpha = \{(x_0, x_1) \in X \times X \mid \text{there exists a winning strategy of } D \text{ for } (x_0, x_1)\}$$

Theorem 6 ([18]). *Assume that F preserves weak pullbacks and has a separating set of monotone evaluation maps. Then $x_0 \sim x_1$ iff D has a winning strategy for the initial configuration (x_0, x_1).*

In the following, we will prove that the greatest fixpoint of $\mathcal{F}_\alpha$ (i.e. $\nu\mathcal{F}_\alpha$) coincides with W_α and hence gives us bisimilarity. Note that $\mathcal{F}_\alpha$ splits classes with respect to only a single equivalence class P. This is different from other coalgebraic partition refinement algorithms where the current equivalence relation is represented by a surjection e with domain X and we separate x_0, x_1 whenever $Fe(\alpha(x_0)) \neq Fe(\alpha(x_1))$, which intuitively means that we split with respect to all equivalence classes at once. Hence we will need to impose extra requirements on the functor, spelled out below, in order to obtain this result.

One direction of the proof deals with deriving a winning strategy for S for each pair $(x_0, x_1) \notin \nu\mathcal{F}_\alpha$. In order to explicitly extract such a winning strategy for S – which will also be important later when we construct the distinguishing formula – we will slightly adapt the algorithm based on fixpoint iteration. Before we come to this, we formally define and explain the strategies of D and S.

We start with the winning strategy of the duplicator in the case where the two given states are bisimilar. This strategy has already been presented in [18], but we describe it here again explicitly. The duplicator only has to know a suitable coalgebra homomorphism.

Proposition 7 (Strategy of the duplicator, [18]). *Let $\alpha\colon X \dashrightarrow FX$ be a coalgebra. Assume that D, S play the game on an initial configuration (x_0, x_1) with $x_0 \sim x_1$. This means that there exists a coalgebra homomorphism $f\colon X \to Z$ from α to a coalgebra $\beta\colon Z \to FZ$ such that $f(x_0) = f(x_1)$.*

Assume that in Step 2 D answers with $p_{1-j} = [p_j]_{ker(f)}$, i.e., p_{1-j} is the $ker(f)$-closure^2 of the predicate p_j. (In other words: $p_{1-j}(s) = 1$ iff there exists $t \in X$ such that $f(s) = f(t)$ and $p_j(t) = 1$).

Then the condition of Step 2 is satisfied and in Step 4 D is always able to pick a state $x'_{1-\ell}$ with $p_{1-\ell}(x'_{1-\ell}) = 1$ and $f(x'_\ell) = f(x'_{1-\ell})$.

We argue why this strategy is actually winning: Since f is a coalgebra homomorphism we have $Ff(\alpha(x_0)) = \beta(f(x_0)) = \beta(f(x_1)) = Ff(\alpha(x_1))$. By construction, p_{1-j} factors through f, that is $p_{1-j} = p'_j \circ f$ for some $p'_j\colon Z \to 2$. This implies $Fp_{1-j}(\alpha(x_j)) = Fp'_j(Ff(\alpha(x_j))) = Fp'_j(Ff(\alpha(x_{1-j}))) = Fp_{1-j}(\alpha(x_{1-j}))$. Since $p_j \leq p_{1-j}$ it follows from monotonicity (Lemma 1) that $Fp_j(\alpha(x_j)) \leq^F Fp_{1-j}(\alpha(x_j)) = Fp_{1-j}(\alpha(x_{1-j}))$. Hence p_{1-j} satisfies the conditions of Step 2. Furthermore if the spoiler picks a state x'_ℓ in p_j in Step 3, the duplicator can pick the same state in p_{1-j} in Step 4. If instead the spoiler picks a state x'_ℓ in p_{1-j}, the duplicator can, due to the closure, at least pick a state $x'_{1-\ell}$ in p_j which satisfies $f(x'_{1-\ell}) = f(x'_\ell)$, which means that the game can continue.

We now switch to the spoiler strategy that can be used to explain why the states are not bisimilar. A strategy for the spoiler is given by a pair of functions

$$I\colon X \times X \to \mathbb{N}_0 \cup \{\infty\} \quad \text{and} \quad T\colon (X \times X)\backslash\nu\mathcal{F}_\alpha \to X \times \mathcal{P}X.$$

Here, $I(x_0, x_1)$ denotes the first index where x_0, x_1 are separated in the fixpoint iteration of $\mathcal{F}_\alpha$. The second component T tells the spoiler what to play in Step 1. In particular whenever $T(x_0, x_1) = (x_j, P)$, S will play j (uniquely determined by x_j unless $x_0 = x_1$, in which case S does not win) and $p_j = \chi_P$.

In the case $I(x_0, x_1) < \infty$ such a winning strategy for S can be computed during fixpoint iteration, see Algorithm 1. Assume that the algorithm terminates after n steps and returns R_n. It is easy to see that R_n coincides with $\nu\mathcal{F}_\alpha$: as usual for partition refinement, we start with the coarsest relation $R_0 = X \times X$. Since $\leq^F$ is, by assumption, anti-symmetric $F\chi_P(\alpha(x_0)) \leq^F F\chi_P(\alpha(x_1))$ and $F\chi_P(\alpha(x_1)) \leq^F F\chi_P(\alpha(x_0))$ are equivalent to $F\chi_P(\alpha(x_0)) = F\chi_P(\alpha(x_1))$ and the algorithm removes a pair (x_0, x_1) from the relation iff this condition does not hold. In addition, $T(x_0, x_1)$ and $I(x_0, x_1)$ are updated, where we distinguish whether $Fp(\alpha(x_0)) \not\leq^F Fp(\alpha(x_1))$ or $Fp(\alpha(x_0)) \not\geq^F Fp(\alpha(x_1))$ hold.

Every relation R_i is finer than its predecessor R_{i-1} and, since $\mathcal{F}_\alpha$ preserves equivalences, each R_i is an equivalence relation. Since we are assuming a finite set X of states, the algorithm will eventually terminate.

We will now show that Algorithm 1 indeed computes a winning strategy for the spoiler.

---

^{2} For a function $f\colon X \to Y$, $ker(f) = \{(x_0, x_1) \mid x_0, x_1 \in X, f(x_0) = f(x_1)\} \subseteq X \times X$.

Algorithm 1. Computation of $\nu \mathcal{F}_\alpha$ and the winning strategy of the spoiler

---

1: **procedure** COMPUTE GREATEST FIXPOINT OF $\mathcal{F}_\alpha$ AND WINNING MOVES FOR S
2:      **for all** $(x_0, x_1) \in X \times X$ **do**
3:         $I(x_0, x_1) \leftarrow \infty$
4:      $i \leftarrow 0, R_0 \leftarrow X \times X$
5:      **repeat**
6:         $i \leftarrow i + 1, R_i \leftarrow R_{i-1}$
7:         **for all** $(x_0, x_1) \in R_{i-1}$ **do**
8:            **for all** $P \in E(R_{i-1})$ **do**
9:               **if** $F\chi_P(\alpha(x_0)) \not\leq^F F\chi_P(\alpha(x_1))$ **then**
10:                 $T(x_0, x_1) \leftarrow (x_0, P), I(x_0, x_1) \leftarrow i, R_i \leftarrow R_i \setminus \{(x_0, x_1)\}$
11:              **else**
12:                 **if** $F\chi_P(\alpha(x_1)) \not\leq^F F\chi_P(\alpha(x_0))$ **then**
13:                    $T(x_0, x_1) \leftarrow (x_1, P), I(x_0, x_1) \leftarrow i, R_i \leftarrow R_i \setminus \{(x_0, x_1)\}$
14:      **until** $R_{i-1} = R_i$
15: **return** R_i, T, I

Proposition 8. *Assume that $R_n = \nu \mathcal{F}_\alpha, T, I$ have been computed by Algorithm 1. Furthermore let $(x_0, x_1) \notin R_n$, which means that $I(x_0, x_1) < \infty$ and $T(x_0, x_1)$ is defined. Then the following constitutes a winning strategy for the spoiler:*

- *Let $T(x_0, x_1) = (x_j, P)$. Then in Step 1 S plays $j \in \{0, 1\}$ and the predicate $p_j = \chi_P$.*
- *Assume that in Step 2, D answers with a state x_{1-j} and a predicate p_{1-j} such that $Fp_j(\alpha(x_j)) \leq^F Fp_{1-j}(\alpha(x_{1-j}))$.*
- *Then, in Step 3 there exists a state $x'_{1-j} \in X$ such that $p_{1-j}(x'_{1-j}) = 1$ and $I(x'_j, x'_{1-j}) < I(x_0, x_1)$ for all $x'_j \in X$ with $p_j(x'_j) = 1$. S will hence select $\ell = 1 - j$, i.e. p_{1-j}, and this state x'_{1-j}.*
- *In Step 4, D selects some x'_j with $p_j(x'_j) = 1$ and the game continues with (x'_0, x'_1) where $(x'_0, x'_1) \in R_n$ and $I(x'_0, x'_1) < I(x_0, x_1)$.*

Finally, we show that $\nu \mathcal{F}_\alpha$ coincides with W_α and therefore also with behavioural equivalence $\sim$ (see [18]). For this purpose, we need one further requirement on the functor:

Definition 9. *Let $F \colon \mathbf{Set} \to \mathbf{Set}$ be an endofunctor on $\mathbf{Set}$. Given a set X, F is separable by singletons on X if the following holds: for all $t_0 \neq t_1$ with $t_0, t_1 \in FX$, there exists $p \colon X \to 2$ where $p(x) = 1$ for exactly one $x \in X$ (i.e., p is a singleton) and $Fp(t_0) \neq Fp(t_1)$. Moreover, F is separable by singletons if F is separable by singletons on all sets X.*

It is obvious that separability by singletons implies the existence of a separating set of predicate liftings, however the reverse implication does not hold as the following example shows.

Example 10. A functor that does not have this property, but does have a separating set of predicate liftings, is the monotone neighbourhood functor $\mathcal{M}$ with $\mathcal{M}X = \{Y \in \mathcal{Q}\mathcal{Q}X \mid Y \text{ upwards-closed}\}$ (see e.g. [12]), where $\mathcal{Q}$ is the contravariant powerset functor. Consider $X = \{a, b, c, d\}$ and $t_0, t_1 \in \mathcal{M}X$ where $t_0 = \uparrow \{\{a, b\}, \{c, d\}\}$, $t_1 = \uparrow \{\{a, b, c\}, \{a, b, d\}, \{c, d\}\}$. That is, the only difference is that t_0 contains the two-element set $\{a, b\}$ and t_1 does not. For any singleton predicate p, the image of $\mathcal{Q}p \colon \mathcal{P}2 \to \mathcal{P}X$ does not contain a two-element set, hence $\mathcal{M}p(t_0) = \mathcal{M}p(t_1)$ – since t_1 and t_2 agree on subsets of X of cardinality different from 2 – and t_0, t_1 cannot be distinguished.

By contrast, both the finite powerset functor $\mathcal{P}_f$ and the finitely supported probability distribution functor $\mathcal{D}$ (which are both ω-accessible and hence yield a logic with only finite formulas) are separable by singletons.

As announced, separability by singletons implies that the fixpoint $\nu\mathcal{F}_\alpha$ coincides with behavioural equivalence:

Theorem 11. *Let F be separable by singletons, and let $\alpha \colon X \to FX$ be an F-coalgebra. Then $\nu\mathcal{F}_\alpha = W_\alpha$, i.e., $\nu\mathcal{F}_\alpha$ contains exactly the pairs $(x_0, x_1) \in X \times X$ for which the duplicator has a winning strategy.*

Example 12. We revisit Example 5 and explain the execution of Algorithm 1. In the first iteration we only have to consider one predicate χ_X, and for all separated pairs of states (s, t) we set $I(s, t) = 1$ where the second component of $T(s, t)$ is X. That is, the states are simply divided into equivalence classes according to their outgoing transitions. More concretely, we obtain the separation of $\{1, 2, 3\}$ (with value $\{(a, 1)\}$) from $\{4\}$ (with value $\{(b, 1)\}$), $\{5\}$ (with value $\{(a, 1), (b, 1)\}$ and $\{6, 7, 8, 9\}$ (with value $\emptyset$). In the second iteration the predicate $\chi_{\{4\}}$ is employed to separate $\{1\}$ (with value $\{(a, 0), (a, 1)\}$) from $\{2\}$ (with value $\{(a, 0)\}$) and we get $I(1, 2) = 2$ with $T(1, 2) = (1, \{4\})$, which also determines the strategy of the spoiler explained above. Similarly $\{3\}$ can be separated from both $\{1\}$ and $\{2\}$ with the predicate $\chi_{\{6,7,8,9\}}$.

The notion of separability by singletons is needed because the partition refinement algorithm we are using separates two states based on a *single* equivalence class of their successors, whereas other partition refinement algorithms (e.g. [17]) consider *all* equivalence classes. As shown in Example 10, this is indeed a restriction, however such additional assumptions seem necessary if we want to adapt efficient bisimulation checking algorithms such as the ones by Kanellakis/Smolka [15] or Paige/Tarjan [22] to the coalgebraic setting. In fact, the Paige/Tarjan algorithm already has a coalgebraic version [9] which operates under the assumption that the functor is *zippable*. Here we show that the related notion of m-zippability is very similar to separability by singletons. (The zippability of [9] is in fact 2-zippability, which is strictly weaker than 3-zippability [29,30].)

Definition 13 (zippability). A functor F is m-*zippable* if the map

$$F(A_1 + \cdots + A_m) \xrightarrow{\langle F(f_1),\ldots,F(f_m)\rangle} F(A_1 + 1) \times \cdots \times F(A_m + 1)$$

is injective for all sets $A_1, \ldots, A_m$, where $f_i = id_{A_i} + \; ! : A_1 + \cdots + A_m \to A_i + 1$, with $! : A_1 + \cdots + A_{i-1} + A_{i+1} + \cdots + A_m \to 1$, is the function mapping all elements of A_i to themselves and all other elements to $\bullet$ (assuming that $1 = \{\bullet\}$).

Lemma 14. *If a functor F is separable by singletons, then F is m-zippable for all m. Conversely, if F is m-zippable, then F is separable by singletons on all sets X with $|X| \le m$.*

Runtime Analysis. We assume that X is finite and that the inequalities in Algorithm 1 (with respect to $\le^F$) are decidable in polynomial time. Then our algorithm terminates and has polynomial runtime.

In fact, if $|X| = n$, the algorithm runs through at most n iterations, since there can be at most n splits of equivalence classes. In each iteration we consider up to n^2 pairs of states, and in order to decide whether a pair can be separated, we have to consider up to n equivalence classes, which results in $O(n^4)$ steps (not counting the steps required to decide the inequalities).

For a finite label set A, the inequalities are decidable in linear time for the functors in our examples ($F = \mathcal{P}_f(A \times (-))$ and $F = (\mathcal{D}(-) + 1)^A$). We expect that we can exploit optimizations based on [15, 22]. In particular one could incorporate the generalization of the Paige-Tarjan algorithm to the coalgebraic setting [9].

4 Construction of Distinguishing Formulas

Next we illustrate how to derive a distinguishing modal formula from the winning strategy of S computed by Algorithm 1. The other direction (obtaining the winning strategy from a distinguishing formula) has been covered in [18].

4.1 Cone Modalities

We focus on an on-the-fly extraction of relevant modalities, to our knowledge a new contribution, and discuss the connection to other – given – sets of separating predicate liftings.

One way of enabling the construction of formulas is to specify the separating set of predicate liftings Λ in advance. But this set might be infinite and hard to represent. Instead here we generate the modalities while constructing the formula. We focus in particular on what we call *cone modalities*: given $v \in F2$ we take the upward-closure of v as a modality.

We also explain how logical formulas with cone modalities can be translated into other separating sets of modalities.

Definition 15 (Cone modalities). Let $v \in F2$. A cone modality $[\uparrow v]$ is given by the following evaluation map $\uparrow v : F2 \to 2$:

$$\uparrow v(u) = \lambda(u) = \begin{cases} 1, & \text{if } v \le^F u \\ 0, & \text{otherwise} \end{cases}$$

Under our running assumptions, these evaluation maps yield a separating set of predicate liftings: Since F has a separating set of monotone predicate liftings, it suffices to show that the evaluation maps are jointly injective on $F2$. Now if $v_0 \neq v_1$ for $v_0, v_1 \in F2$, then w.l.o.g. $v_0 \not\leq^F v_1$, since we require that the lifted order is anti-symmetric. Hence, $\uparrow v_0(v_0) = 1$ and $\uparrow v_0(v_1) = 0$.

Example 16. We discuss modalities respectively evaluation maps in more detail for the functor $F = \mathcal{P}_f(A \times (-))$ (see also Example 5). In our example, $A = \{a, b\}$. The set $F2$ with order $\leq^F$ is depicted as a Hasse diagram in Fig. 2. For every element there is a cone modality, 16 modalities in total. It is known from the Hennessy-Milner theorem [14] that two modalities are enough: either $\Box_a, \Box_b$ (box modalities) or $\Diamond_a, \Diamond_b$ (diamond modalities), where for $v \in F2$,

$$\Box_a(v) = \begin{cases} 1 & \text{if } (a,0) \notin v \\ 0 & \text{otherwise} \end{cases} \qquad \Diamond_a(v) = \begin{cases} 1 & \text{if } (a,1) \in v \\ 0 & \text{otherwise.} \end{cases}$$

In Fig. 2, $\Box_a$ respectively $\Diamond_a$ are represented by the elements above the two lines (solid respectively dashed).

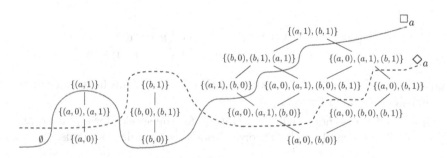

Fig. 2. The set $F2 = \mathcal{P}_f(\{a,b\} \times 2)$ with the order $\leq^F$ (for labelled transition systems). The modality $\Box_a$ ($\Diamond_a$) is given by the elements above the solid (dashed) line.

Example 17. As a second example we discuss the functor $F = (\mathcal{D}(-) + 1)^A$, specifying probabilistic transition systems. The singleton set $1 = \{\bullet\}$ denotes termination. Again we set $A = \{a, b\}$.

Since $\mathcal{D}2$ is isomorphic to the interval $[0, 1]$, we can simply represent any distribution $d : 2 \to [0, 1]$ by $d(1)$. Hence $F2 \cong ([0, 1] + 1)^A$. The partial order is componentwise and is depicted in Fig. 3: it decomposes into four disjoint partial orders, depending on which of a, b are mapped to $\bullet$. The right-hand part of this partial order consists of function $[0, 1]^A$ with the pointwise order.

We will also abbreviate a map $[a \mapsto p, b \mapsto q]$ by $\langle a_p, b_q \rangle$.

$$[a \mapsto 1, b \mapsto \bullet] \qquad [a \mapsto \bullet, b \mapsto 1] \qquad \begin{array}{c} [a \mapsto 1, b \mapsto 1] \\ \vdots \\ \end{array}$$

$$[a \mapsto \bullet, b \mapsto \bullet] \qquad [a \mapsto 0, b \mapsto \bullet] \qquad [a \mapsto \bullet, b \mapsto 0] \qquad [a \mapsto 0, b \mapsto 0]$$

Fig. 3. $F2 \cong ([0,1]+1)^A$ with order $\leq^F$ (for probabilistic transition systems).

4.2 From Winning Strategies to Distinguishing Formulas

We will now show how a winning strategy of S can be transformed into a distinguishing formula, based on cone modalities, including some examples.

The basic idea behind the construction in Definition 18 is the following: Let (x_0, x_1) be a pair of states separated in the i-th iteration of the partition refinement algorithm (Algorithm 1). This means that we have the following situation: $F\chi_P(\alpha(x_0)) \not\leq^F F\chi_P(\alpha(x_1))$ (or vice versa) for some equivalence class P of R_{i-1}. Based on $v = F\chi_P(\alpha(x_0))$ we define a cone modality $\lambda = \uparrow v$. Now, if we can characterize P by some formula ψ, i.e., $[\![\psi]\!] = \chi_P$ (we will later show that this is always possible), we can define the formula $\varphi = [\lambda]\psi$. Then it holds that:

$$[\![\varphi]\!](x_0) = \lambda(F[\![\psi]\!](\alpha(x_0))) = \uparrow v(F\chi_P(\alpha(x_0))) = 1$$
$$[\![\varphi]\!](x_1) = \lambda(F[\![\psi]\!](\alpha(x_1))) = \uparrow v(F\chi_P(\alpha(x_1))) = 0$$

That is we have $x_0 \models \varphi$ and $x_1 \not\models \varphi$, which means that we have constructed a distinguishing formula for x_0, x_1.

First, we describe how a winning strategy for the spoiler for a pair (x_0, x_1) is converted into a formula and then prove that this formula distinguishes x_0, x_1.

Definition 18. Let $x_0 \nsim x_1$ (equivalently $(x_0, x_1) \notin R_n$), and let (T, I) be the winning strategy for the spoiler computed by Algorithm 1. We construct a formula φ_{x_0, x_1} as follows: assume that $T(x_0, x_1) = (s, P)$ where $s = x_0$. Then set $v = F\chi_P(\alpha(x_0))$, $\lambda = \uparrow v$ and $\varphi_{x_0, x_1} = [\lambda]\varphi$, where φ is constructed by recursion as follows:

– $I(x_0, x_1) = 1$: $\varphi = tt$
– $I(x_0, x_1) > 1$: $\varphi = \bigvee_{x_0' \in P} \left(\bigwedge_{x_1' \in X \setminus P} \varphi_{x_0', x_1'} \right)$

If $s = x_1$, then we set $v = F\chi_P(\alpha(x_1))$ and $\varphi_{x_0, x_1} = \neg[\lambda]\varphi$ instead. The recursion terminates because $I(x_0', x_1') < I(x_0, x_1)$ (since P is an equivalence class of R_{i-1} where $i = I(x_0, x_1)$).

Proposition 19. *Let $\alpha : X \to FX$ be a coalgebra and assume that we have computed R_n, T, I with Algorithm 1. Then, given $(x_0, x_1) \notin R_n$, the construction in Definition 18 yields a formula $\varphi_{x_0, x_1} \in \mathcal{L}^\kappa(\Lambda)$ such that $x_0 \models \varphi_{x_0, x_1}$ and $x_1 \not\models \varphi_{x_0, x_1}$.*

We next present an optimization of the construction in Definition 18, inspired by [6]. In the case $I(x_0, x_1) > 1$ one can pick an arbitrary $x_0' \in P$ and keep only one element of the disjunction.

In order to show that this simplification is permissible, we need the following lemma.

Lemma 20. *Given two states* $(x_0, x_1) \notin R_n$ *and a distinguishing formula* φ_{x_0, x_1} *based on Definition 18. Let* (x_0', x_1') *be given such that* $I(x_0', x_1') > I(x_0, x_1)$. *Then* $x_0' \vDash \varphi_{x_0, x_1}$ *if and only if* $x_1' \vDash \varphi_{x_0, x_1}$.

Now we can show that we can replace the formula φ from Definition 18 by a simpler formula φ'.

Lemma 21. *Let* $(x_0, x_1) \notin R_i$ *and let* P *be an equivalence class of* R_{i-1}. *Furthermore let*

$$\varphi' = \bigwedge_{x_1' \in X \backslash P} \varphi_{x_0', x_1'}$$

for some $x_0' \in P$. *Then* $\llbracket \varphi' \rrbracket = \chi_P$.

Finally, we can simplify our construction described in Definition 18 to only one inner conjunction.

Corollary 22. *We use the construction of* φ_{x_0, x_1} *as described in Definition 18 with the only modification that for* $I(x_0, x_1) > 1$ *the formula* φ *is replaced by*

$$\varphi' = \bigwedge_{x_1' \in X \backslash P} \varphi_{x_0', x_1'}$$

for some $x_0' \in P$. *Then this yields a formula* φ_{x_0, x_1} *such that* $x_0 \vDash \varphi_{x_0, x_1}$ *and* $x_1 \nvDash \varphi_{x_0, x_1}$.

A further optimization takes only one representative x_1' from every equivalence class different from P.

We now explore two slightly more complex examples.

Example 23. Take the coalgebra for the functor $F = (\mathcal{D}(-) + 1)^A$ depicted in Fig. 4, with $A = \{a, b\}$ and set $X = \{1, \ldots, 5\}$ of states. For instance, $\alpha(3) = [a \mapsto \delta_3, b \mapsto \bullet]$ where δ_3 is the Dirac distribution. This is visualized by drawing an arrow labelled $a, 1$ from 3 to 3 and omitting b-labelled arrows.

We explain only selected steps of the construction: In the first step, the partition refinement algorithm (Algorithm 1) separates 1 from 3 (among other separations), where the spoiler strategy is given by $T(1, 3) = (1, X)$. In order to obtain a distinguishing formula, we determine $v = F\chi_X(\alpha(1)) = \langle a_1, b_1 \rangle$ (using the abbreviations explained in Example 17) and obtain $\varphi_{1,3} = [\uparrow \langle a_1, b_1 \rangle] tt$. In fact, this formula also distinguishes 1 from 4, hence $\varphi_{1,3} = \varphi_{1,4}$. If, on the other hand, we want to distinguish 3, 4, we obtain $\varphi_{3,4} = [\uparrow \langle a_1, b_\bullet \rangle] tt$.

After the first iteration, we obtain the partition $\{1, 2, 5\}, \{3\}, \{4\}$. Now we consider states 1, 2 which can be separated by playing $T(2, 1) = (2, \{1, 2, 5\})$, since 5 behaves differently from 3. Again we compute $v = F\chi_P(\alpha(2)) = \langle a_1, b_{0.8}\rangle$ (for $P = \{1, 2, 5\}$) and obtain $\varphi_{2,1} = [\uparrow\langle a_1, b_{0.8}\rangle](\varphi_{1,3} \wedge \varphi_{1,4})$. Here we picked 1 as the representative of its equivalence class.

In summary we obtain $\varphi_{2,1} = [\uparrow\langle a_1, b_{0.8}\rangle][\uparrow\langle a_1, b_1\rangle]tt$, which is satisfied by 2 but not by 1.

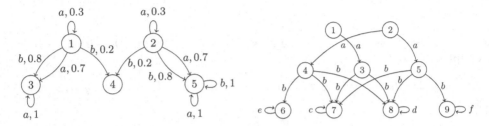

Fig. 4. Probabilistic transition system

Fig. 5. Non-deterministic transition system.

Example 24. We will now give an example where conjunction is required to obtain the distinguishing formula. We work with the coalgebra for the functor $F = \mathcal{P}_f(A \times (-))$ depicted in Fig. 5, with $A = \{a, b, c, d, e, f\}$ and set $X = \{1, \ldots, 9\}$ of states.

We explain only selected steps: In the first step, the partition refinement separates 6 from 7 (among other separations), where the spoiler strategy is given by $T(6, 7) = (6, X)$. As explained above, we determine $v = F\chi_X(\alpha(6)) = \{(e, 1)\}$ and obtain $\varphi_{6,7} = [\uparrow\{(e, 1)\}]tt$. In fact, this formula also distinguishes 6 from all other states, so we denote it by $\varphi_{6,*}$.

Next, we consider the states $3, 4$, where the possible moves of 3 are a proper subset of the moves of 4. Hence the spoiler strategy is $T(3, 4) = (4, \{6\})$, i.e., the spoiler has to move to state 6, which is not reachable from 3. Again we compute $v = F\chi_P(\alpha(4)) = \{(b, 1), (b, 0)\}$ (for $P = \{6\}$) and obtain $\varphi_{3,4} = \neg[\uparrow \{(b, 1), (b, 0)\}]\varphi_{6,*}$. Note that this time we have to use negation, since the spoiler moves from the second state in the pair.

Finally, we consider the states $1, 2$, where the spoiler strategy is $T(1, 2) = (1, \{3\})$. We compute $v = F\chi_P(\alpha(1)) = \{(a, 1)\}$ (for $P = \{3\}$) and obtain $\varphi_{1,2} = [\uparrow\{(a, 1)\}](\bigwedge_{x\in\{1,2,4,\ldots,9\}} \varphi_{3,x})$. In fact, here it is sufficient to consider $x = 4$ and $x = 5$, resulting in the following distinguishing formula:

$$[\uparrow\{(a, 1)\}](\neg[\uparrow\{(b, 0), (b, 1)\}][\uparrow\{(e, 1)\}]tt \wedge \neg[\uparrow\{(b, 0), (b, 1)\}][\uparrow\{(f, 1)\}]tt).$$

4.3   Recoding Modalities

Finally, we will show under which conditions one can encode cone modalities into given generic modalities, determined by a separating set of predicate liftings Λ, not necessarily monotone. We first need the notion of strong separation.

Definition 25. Let Λ be a separating set of predicate liftings of the form $\lambda \colon F2 \to 2$. We call Λ *strongly separating* if for every $t_0 \neq t_1$ with $t_0, t_1 \in F2$ there exists $\lambda \in \Lambda$ such that $\lambda(t_0) \neq \lambda(t_1)$.

We can generate a set of strongly separating predicate liftings from every separating set of predicate liftings.

Lemma 26. *Let Λ be a separating set of predicate liftings. Furthermore we denote the four functions on 2 by id_2, one (constant 1-function), zero (constant 0-function) and neg ($neg(0) = 1$, $neg(1) = 0$).*
Then
$$\Lambda' = \{\lambda, \lambda \circ F\,one, \lambda \circ F\,zero, \lambda \circ F\,neg \mid \lambda \in \Lambda\}$$
is a set of strongly separating predicate liftings.
Furthermore for every formula φ we have that

$$[\lambda \circ F\,one]\varphi \equiv [\lambda]tt \qquad [\lambda \circ F\,zero]\varphi \equiv [\lambda]f\!f \qquad [\lambda \circ F\,neg]\varphi \equiv [\lambda](\neg\varphi)$$

This means that we can still express the new modalities with the previous ones. Λ' is just an auxiliary construct that helps us to state the following proposition. The construction of Λ' from Λ was already considered in [26, Definition 24], where it is called *closure*.

Proposition 27. *Suppose that $F2$ is finite, and let Λ be a strongly separating set of predicate liftings. Moreover, let $v \in F2$, and let φ be a formula. For $u \in F2$, we write $\Lambda_u = \{\lambda \in \Lambda \mid \lambda(u) = 1\}$. Then*

$$[\uparrow v]\varphi \equiv \bigvee_{v \leq^F u} \left(\bigwedge_{\lambda \in \Lambda_u} [\lambda]\varphi \wedge \bigwedge_{\lambda \notin \Lambda_u} \neg[\lambda]\varphi \right).$$

By performing this encoding inductively, we can transform a formula with cone modalities into a formula with modalities in Λ. The encoding preserves negation and conjunction, only the modalities are transformed.

Example 28. We come back to labelled transition systems and the functor $F = \mathcal{P}_f(A \times (-))$, with $A = \{a, b\}$. In this case the set $\{\Box_a, \Box_b, \Diamond_a, \Diamond_b\}$ of predicate liftings is strongly separating.

Now let $v = \{(a, 0), (b, 1)\} \in \mathcal{P}_f(A \times 2)$. We show how to encode the corresponding cone modality using only box and diamond:

$$[\uparrow v]\varphi \equiv (\neg\Box_a\varphi \wedge \Box_b\varphi \wedge \neg\Diamond_a\varphi \wedge \Diamond_b\varphi) \vee (\neg\Box_a\varphi \wedge \Box_b\varphi \wedge \Diamond_a\varphi \wedge \Diamond_b\varphi)$$
$$\vee (\Box_a\varphi \wedge \Box_b\varphi \wedge \Diamond_a\varphi \wedge \Diamond_b\varphi)$$

The first term describes $\{(a, 0), (b, 1)\}$, the second $\{(a, 0), (a, 1), (b, 1)\}$ and the third $\{(a, 1), (b, 1)\}$.

Note that we cannot directly generalize Proposition 27 to the case where $F2$ is infinite. The reason for this is that the disjunction over all $u \in F2$ such that $v \leq^F u$ might violate the cardinality constraints of the logic. Hence we will consider an alternative, where the re-coding works only under certain assumptions. We will start with the following example.

Example 29. Consider the functor $F = (\mathcal{D}(-)+1)^A$ (see also Example 17) and the corresponding (countable) separating set of (monotone) predicate liftings

$$\Lambda = \{\lambda_{(a,q)} : F2 \to 2 \mid a \in A, q \in [0,1] \cap \mathbb{Q}\} \cup \{\lambda_{(a,\bullet)} \mid a \in A\}$$

where $\lambda_{(a,q)}(v) = 1$ if $v(a) \in \mathbb{R}$ and $v(a) \geq q$ and $\lambda_{(a,\bullet)} = 1$ if $v(a) = \bullet$. Here, $[\lambda_{(a,q)}]\varphi$ indicates that we do not terminate with a, and the probability of reaching a state satisfying φ under an a-transition is at least q, and a modality $[\lambda_{(a,\bullet)}]$ ignores its argument formula, and tells us that we terminate with a.

The disjunction $\bigvee_{v \leq^F u}$ in the construction of $[\uparrow v]\varphi$ in Proposition 27 is in general uncountable and may hence fail to satisfy the cardinality constraints of the logic. However, we can exploit certain properties of this set of predicate liftings, in order to re-code modalities.

Lemma 30. *Let F be the functor with $F = (\mathcal{D}(-) + 1)^A$ and let Λ be the separating set of predicate liftings from Example 29. Then*

$$\uparrow v = \bigcap_{\lambda \in \Lambda, \lambda(v)=1} \lambda \quad \text{for all } v \in F2. \tag{1}$$

Note that this property does not hold for the $\Box$ and $\Diamond$ modalities for the functor $F = \mathcal{P}_f(A \times (-))$. This can be seen via Fig. 2, where the upward closure of $\{(b,0)\}$ contains three elements. However, $\{(b,0)\}$ is only contained in the modality $\Box_a$ (and no other modality), which does not coincide with the upward-closure of $\{(b,0)\}$.

The following proposition, which relates to the well-known fact that predicate liftings are closed under infinitary Boolean combinations (e.g. [26]), provides a recipe for transforming cone modalities $\uparrow v$ into given modalities Λ satisfying (1) as in Lemma 30:

Proposition 31. *Given a set $\Lambda' \subseteq \Lambda$ of predicate liftings, understood as subsets of $F2$, we have*

$$[\bigcap_{\lambda \in \Lambda'} \lambda]\varphi \equiv \bigwedge_{\lambda \in \Lambda'} [\lambda]\varphi.$$

Note that this construction might again violate the cardinality constraints of the logic. In particular, for the probabilistic case (Example 17) we have finite formulas, but countably many modalities. However, if we assume that the set of labels A is finite and restrict the coefficients in the coalgebra to rational numbers, every cone modality can be represented as the intersection of only finitely many minimal given modalities and so the encoding preserves finiteness.

5   T-Beg: A Generic Tool for Games and the Construction of Distinguishing Formulas

5.1   Overview

A tool for playing bisimulation games is useful for teaching, for illustrating examples in talks, for case studies and in general for interaction with the user. There are already available tools, providing visual feedback to help the user understand why two states are (not) bisimilar, such as THE BISIMULATION GAME GAME[3] or BISIMULATION GAMES TOOLS[4] [10]. Both games are designed for labelled transition systems and [10] also covers branching bisimulation.

Our tool T-BEG goes beyond labelled transition system and allows to treat coalgebras in general (under the restrictions that we impose), that is, we exploit the categorical view to create a generic tool. As shown earlier in Sects. 3 and 4, the coalgebraic game defined in Definition 2 provides us with a generic algorithm to compute the winning strategies and distinguishing formulas.

The user can either take on the role of the spoiler or of the duplicator, playing on some coalgebra against the computer. The tool computes the winning strategy (if any) and follows this winning strategy if possible. We have also implemented the construction of the distinguishing formula for two non-bisimilar states.

The genericity over the functor is in practice achieved as follows: The user either selects an existing functor F (e.g. the running examples of the paper), or implements his/her own functor by providing the code of one class with nine methods (explained below). Everything else, such as embedding the functor into the game and the visualization are automatically handled by T-BEG. In the case of weighted systems, T-BEG even handles the graphical representation.

Then, he/she enters or loads a coalgebra $\alpha : X \to FX$ (with X finite), stored as *csv* (comma separated value) file. Now the user can switch to the game view and start the game by choosing one of the two roles (spoiler or duplicator) and selecting a pair of states (x_0, x_1), based on the visual graph representation.

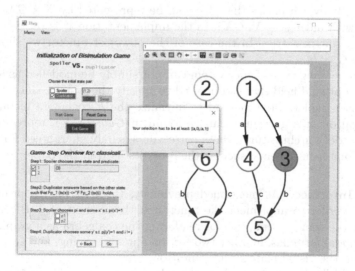

Fig. 6. Screenshot of the graphical user interface with a game being played. (Color figure online)

---

[3] http://www.brics.dk/bisim/.

[4] https://www.jeroenkeiren.nl/blog/on-games-and-simulations/.

Next, the computer takes over the remaining role and the game starts: In the game overview, the user is guided through the steps by using two colors to indicate whether it is spoiler's (violet) or duplicator's (cyan) turn (see Fig. 6).

In the case of two non-bisimular states, the tool will display a distinguishing formula at the end of the game.

5.2 Design

We now give an overview over the design and the relevant methods within the tool. We will also explain what has to be done in order to integrate a new functor.

T-BEG is a Windows tool offering a complete graphical interface, developed in Microsoft's Visual Studio using $C\#$, especially *Generics*. It uses a graph library[5], which in turn provides a *GraphEditor* that allows for storing graphs as *MSAGL* files or as *png* and *jpg* files.

The program is divided into five components: Model, View, Controller, Game and Functor. We have chosen *MVC (Model View Controller)* as a modular pattern, so modules can be exchanged. Here we have several $Model\langle T\rangle$ managed by the *Controller*, where the functor in the sense of a *Functor* class, which always implements the *Functor Interface*, is indicated by the parameter $\langle T\rangle$.

While the tool supports more general functors, there is specific support for functors F with $F = V^{G(-)}$ where V specifies a semiring and G preserves finite sets. That is, F describes the branching type of a weighted transition system, where for instance $G = A \times (-) + 1$ (introducing finitely many labels and termination). Coalgebras are of the form $X \to V^{GX}$ or – via currying – of the form $X \times GX \to V$, which means that they can be represented by $X \times GX$-matrices (matrices with index sets X, GX). In the implementation V is the generic data type of the matrix entries. In the case of the powerset functor we simply have $V = 2$ and $G = Id$.

If the branching type of the system can not simply be modelled as a matrix, there is an optional field that can be used to specify the system, since $Model\langle T\rangle$ calls the user-implemented method to initialize the F-coalgebra instance. The implementation of Algorithm 1 can be found in $Game\langle T,V\rangle$, representing the core of the tool's architecture, whose correctness is only guaranteed for functors that meet our requirements, such as the functors used in the paper.

Functor Interface. As mentioned previously, the user has to provide nine methods in order to implement the functor in the context of T-BEG: two are needed for the computation, two for rendering the coalgebra as a graph, one for creating modal formulas, another two for loading and saving, and two more for customizing the visual matrix representation.

We would like to emphasize here that the user is free to formally implement the functor in the sense of the categorical definition as long as the nine methods needed for the game are provided. In particular, we do not need the application of the functor to arrows since we only need to lift predicates $p : X \to 2$.

---

[5] https://www.nuget.org/packages/Microsoft.Msagl.GraphViewerGDI.

Within *MyFunctor*, which implements the interface $Functor\langle F, V\rangle$, the user defines the data structure F for the branching type of the transition system (e.g., a list or bit vector for the powerset functor, or the corresponding function type in the case of the distribution functor). Further, the user specifies the type V that is needed to define the entries of $X \times GX$ (e.g. a double value for a weight or $0, 1$ to indicate the existence of a transition).

Here we focus on five methods which have to be provided, omitting the remaining four which are less central.

$Matrix\langle F, V\rangle InitMatrix(\dots)$: This method initializes the transition system with the string-based input of the user. The information about the states and the alphabet is provided via an input mask in the form of a matrix.

bool CheckDuplicatorsConditionStep2$(\dots)$: given two states x_0, x_1 and two predicates p_0, p_1, this method checks whether

$$Fp_0(\alpha(x_0)) \leq^F Fp_1(\alpha(x_1)).$$

This method is used when playing the game (in Step 2) and in the partition refinement algorithm (Algorithm 1) for the case $p_0 = p_1$.

TSToGraph$(\dots)$: This method handles the implementation of the graph-based visualization of the transition system. For weighted systems the user can rely on the default implementation included within the *Model*. In this case, arrows between states and their labels are generated automatically.

GraphToTS$(\dots)$: This method is used for the other direction, i.e. to derive the transition system from a directed graph given by *Graph*.

string GetModalityToString$(\dots)$: This method is essential for the automatic generation of the modal logical formulas distinguishing two non-bisimilar states as described in Definition 18. In each call, the cone modality that results from $F\chi_P(\alpha(s))$ with $T(x_0, x_1) = (s, P)$ is converted into a string.

The implementation costs arising on the user side can be improved by employing a separate module that automatically generates functors (see [7]). But it is not clear whether the lifting of the preorder can be obtained automatically.

6   Conclusion and Discussion

Our aim in this paper is to give concrete recipes for explaining non-bisimilarity in a coalgebraic setting. This involves the computation of the winning strategy of the spoiler in the bisimulation game, based on a partition refinement algorithm, as well as the generation of distinguishing formulas, following the ideas of [6]. Furthermore we have presented a tool that implements this functionality in a generic way. Related tools, as mentioned in [10], are limited to labelled transition systems and mainly focus on the spoiler strategy instead of generating distinguishing formulas.

In the future we would like to combine our prototype implementation with an efficient coalgebraic partition refinement algorithm, adapting the ideas of

Kanellakis/Smolka [15] or Paige/Tarjan [22] or using the existing coalgebraic generalization [9], thus enabling the efficient computation of winning strategies and distinguishing formulae.

For the generation of distinguishing formulas, an option would be to fix the modalities a priori and to use them in the game, similar to the notion of λ-bisimulation [11,18]. However, there might be infinitely many modalities and the partition refinement algorithm can not iterate over all of them. A possible solution would be to find a way to check the conditions symbolically in order to obtain suitable modalities.

Of course we are also interested in whether we can lift the extra assumptions that were necessary in order to re-code modalities in Sect. 4.3. We also expect that the bisimulation game can be extended to polyadic predicate liftings.

An interesting further idea is to translate the coalgebra into multi-neighbourhood frames [13,20], based on the predicate liftings, and to derive a λ-bisimulation game as in [11,18] from there. (The λ-bisimulation game does not require weak pullback preservation and extends the class of admissible functors, but requires us to fix the modalities rather than generate them.) One could go on and translate these multi-neighbourhood frames into Kripke frames, but this step unfortunately does not preserve bisimilarity.

We also plan to study applications where we can exploit the fact that the distinguishing formula witnesses non-bisimilarity. For instance, we see interesting uses in the area of differential privacy [5], for which we would need to generalize the theory to a quantitative setting. That is, we would like to construct distinguishing formulas in the setting of quantitative coalgebraic logics, which characterize behavioural distances.

Acknowledgements. We would like to thank the reviewers for the careful reading of the paper and their valuable comments. Furthermore we thank Thorsten Wißmann and Sebastian Küpper for inspiring discussions on (efficient) coalgebraic partition refinement and zippability.

References

1. Balan, A., Kurz, A.: Finitary functors: from Set to Preord and Poset. In: Corradini, A., Klin, B., Cîrstea, C. (eds.) CALCO 2011. LNCS, vol. 6859, pp. 85–99. Springer, Heidelberg (2011). https://doi.org/10.1007/978-3-642-22944-2_7
2. Baltag, A.: Truth-as-simulation: towards a coalgebraic perspective on logic and games. Technical report SEN-R9923, CWI, November 1999
3. Baltag, A.: A logic for coalgebraic simulation. In: Coalgebraic Methods in Computer Science, CMCS 2000. ENTCS, vol. 33, pp. 42–60. Elsevier (2000)
4. Barr, M.: Relational algebras. In: MacLane, S., et al. (eds.) Reports of the Midwest Category Seminar IV. Lecture Notes in Mathematics, vol. 137, pp. 39–55. Springer, Heidelberg (1970). https://doi.org/10.1007/BFb0060439
5. Chatzikokolakis, K., Gebler, D., Palamidessi, C., Xu, L.: Generalized bisimulation metrics. In: Baldan, P., Gorla, D. (eds.) CONCUR 2014. LNCS, vol. 8704, pp. 32–46. Springer, Heidelberg (2014). https://doi.org/10.1007/978-3-662-44584-6_4

6. Cleveland, R.: On automatically explaining bisimulation inequivalence. In: Clarke, E.M., Kurshan, R.P. (eds.) CAV 1990. LNCS, vol. 531, pp. 364–372. Springer, Heidelberg (1991). https://doi.org/10.1007/BFb0023750

7. Deifel, H.-P., Milius, S., Schröder, L., Wißmann, T.: Generic partition refinement and weighted tree automata. In: ter Beek, M.H., McIver, A., Oliveira, J.N. (eds.) FM 2019. LNCS, vol. 11800, pp. 280–297. Springer, Cham (2019). https://doi.org/10.1007/978-3-030-30942-8_18

8. Desharnais, J., Laviolette, F., Tracol, M.: Approximate analysis of probabilistic processes: logic, simulation and games. In: Proceedings of QEST 2008, pp. 264–273. IEEE (2008)

9. Dorsch, U., Milius, S., Schröder, L., Wißmann, T.: Efficient coalgebraic partition refinement. In: Proceedings of CONCUR 2017. LIPIcs, Schloss Dagstuhl - Leibniz-Zentrum fuer Informatik (2017)

10. de FrutosEscrig, D., Keiren, J.J.A., Willemse, T.A.C.: Games for bisimulations and abstraction (2016). https://arxiv.org/abs/1611.00401, arXiv:1611.00401

11. Gorín, D., Schröder, L.: Simulations and bisimulations for coalgebraic modal logics. In: Heckel, R., Milius, S. (eds.) CALCO 2013. LNCS, vol. 8089, pp. 253–266. Springer, Heidelberg (2013). https://doi.org/10.1007/978-3-642-40206-7_19

12. Hansen, H., Kupke, C.: A coalgebraic perspective on monotone modal logic. In: Coalgebraic Methods in Computer Science, CMCS 2004. LNCS, vol. 106, pp. 121–143. Elsevier (2004)

13. Hansen, H.H.: Monotonic modal logics. Master's thesis, University of Amsterdam (2003)

14. Hennessy, M., Milner, A.: Algebraic laws for nondeterminism and concurrency. J. ACM 32(1), 137–161 (1985)

15. Kanellakis, P.C., Smolka, S.A.: CCS expressions, finite state processes, and three problems of equivalence. Inf. Comput. 86, 43–68 (1990)

16. Komorida, Y., Katsumata, S.Y., Hu, N., Klin, B., Hasuo, I.: Codensity games for bisimilarity. In: Proceedings of LICS 2019, pp. 1–13. ACM (2019)

17. König, B., Küpper, S.: A generalized partition refinement algorithm, instantiated to language equivalence checking for weighted automata. Soft Comput. 22(4), 1103–1120 (2018). https://doi.org/10.1007/s00500-016-2363-z

18. König, B., Mika-Michalski, C.: (Metric) Bisimulation games and real-valued modal logics for coalgebras. In: Proceedings of CONCUR 2018. LIPIcs, vol. 118, pp. 37:1–37:17. Schloss Dagstuhl - Leibniz Center for Informatics (2018)

19. König, B., Mika-Michalski, C., Schröder, L.: Explaining non-bisimilarity in a coalgebraic approach: games and distinguishing formulas (2020). https://arxiv.org/abs/2002.11459, arXiv:2002.11459

20. Kracht, M., Wolter, F.: Normal monomodal logics can simulate all others. J. Symb. Log. 64(1), 99–138 (1999)

21. Kupke, C.: Terminal sequence induction via games. In: Bosch, P., Gabelaia, D., Lang, J. (eds.) TbiLLC 2007. LNCS (LNAI), vol. 5422, pp. 257–271. Springer, Heidelberg (2009). https://doi.org/10.1007/978-3-642-00665-4_21

22. Paige, R., Tarjan, R.E.: Three partition refinement algorithms. SIAM J. Comput. 16(6), 973–989 (1987)

23. Pattinson, D.: Coalgebraic modal logic: soundness, completeness and decidability of local consequence. Theoret. Comput. Sci. 309(1), 177–193 (2003)

24. Pattinson, D.: Expressive logics for coalgebras via terminal sequence induction. Notre Dame J. Form. Log. 45, 19–33 (2004)

25. Rutten, J.: Universal coalgebra: a theory of systems. Theoret. Comput. Sci. 249(1), 3–80 (2000)

26. Schröder, L.: Expressivity of coalgebraic modal logic: the limits and beyond. Theoret. Comput. Sci. **390**(2), 230–247 (2008)
27. Stirling, C.: Bisimulation, modal logic and model checking games. Log. J. IGPL **7**(1), 103–124 (1999)
28. Trnková, V.: General theory of relational automata. Fundam. Inform. **3**, 189–234 (1980)
29. Wißmann, T.: Personal communication
30. Wißmann, T.: Coalgebraic semantics and minimization in sets and beyond. Ph.D. thesis, Friedrich-Alexander-Universität Erlangen-Nürnberg (2020)

A Categorical Approach to Secure Compilation

Stelios Tsampas[1], Andreas Nuyts[1], Dominique Devriese[2](✉),
and Frank Piessens[1]

[1] KU Leuven, Leuven, Belgium
{stelios.tsampas,andreas.nuyts,frank.piessens}@cs.kuleuven.be
[2] Vrije Universiteit Brussel, Brussels, Belgium
dominique.devriese@vub.be

Abstract. We introduce a novel approach to secure compilation based
on maps of distributive laws. We demonstrate through four examples that
the coherence criterion for maps of distributive laws can potentially be a
viable alternative for compiler security instead of full abstraction, which
is the preservation and reflection of contextual equivalence. To that end,
we also make use of the well-behavedness properties of distributive laws
to construct a categorical argument for the contextual connotations of
bisimilarity.

Keywords: Secure compilation · Distributive laws · Structural
operational semantics

1 Introduction

As a field, secure compilation is the study of compilers that formally preserve
abstractions across languages. Its roots can be tracked back to the seminal observation made by Abadi [1], namely that compilers which do not protect high-level
abstractions against low-level contexts might introduce security vulnerabilities.
But it was the advent of secure architectures like the Intel SGX [15] and an ever-
increasing need for computer security that motivated researchers to eventually
work on formally proving compiler security.

The most prominent [16,18,32,35,37,45,49] formal criterion for compiler
security is *full abstraction*: A compiler is fully abstract if it preserves and reflects
Morris-style contextual equivalence [31], i.e. indistinguishability under all program contexts, which are usually defined as programs with a hole. The intuition
is that contexts represent the ways an attacker can interact with programs and so
full abstraction ensures that such interactions are consistent between languages.

Full abstraction is arguably a strong and useful property but it is also notori-
ously hard to prove for realistic compilers, mainly due to the inherent challenge
of having to reason directly about program contexts [9,18,24,37]. There is thus a
need for better formal methods, a view shared in the scientific community [10,33].
While recent work has proposed generalizing from full abstraction towards the

© IFIP International Federation for Information Processing 2020
Published by Springer Nature Switzerland AG 2020
D. Petrişan and J. Rot (Eds.): CMCS 2020, LNCS 12094, pp. 155–179, 2020.
https://doi.org/10.1007/978-3-030-57201-3_9

so-called *robust* properties [2,36], the main challenge of quantifying over program contexts remains, which manifests when directly translating target contexts to the source (*back-translation*). Other techniques, such as trace semantics [35] or logical relations [17], require complex correctness and completeness proofs w.r.t. contextual equivalence in order to be applicable.

In this paper we introduce a novel, categorical approach to secure compilation. The approach has two main components: the elegant representation of Structural Operational Semantics (SOS) [38] using category-theoretic *distributive laws* [48][1] and also *maps of distributive laws* [27,40,50] as secure compilers that preserve bisimilarity. Our method aims to be unifying, in that there is a general, shared formalism for operational semantics, and simplifying, in that the formal criterion for compiler security, the *coherence criterion* for maps of distributive laws, is straightforward and relatively easy to prove.

The starting point of our contributions is an abstract proof on how coalgebraic bisimilarity under distributive laws holds *contextual* meaning in a manner similar to contextual equivalence (Sect. 4.3). We argue that this justifies the use of the coherence criterion for testing compiler security as long as bisimilarity adequately captures the underlying threat model. We then demonstrate the effectiveness of our approach by appeal to four examples of compiler (in)security. The examples model classic, non-trivial problems in secure compilation:

- An example of an extra processor register in the target language that conveys additional information about computations (Sect. 5).
- A datatype mismatch between the type of variable (Sect. 6).
- The introduction of illicit control flow in the target language (Sect. 7).
- A case of incorrect local state encapsulation (Sect. 8).

For each of these examples we present an insecure compiler that fails the coherence criterion, then introduce *security primitives* in the target language and construct a secure compiler that respects it. We also examine how bisimilarity can be both a blessing and a curse as its strictness and rigidity sometimes lead to relatively contrived solutions. Finally, in Sect. 9, we discuss related work and point out potential avenues for further development of the underlying theory.

On the Structure and Style of the Paper. This work is presented mainly in the style of programming language semantics but its ideas are deeply rooted in category theory. We follow an "on-demand" approach when it comes to important categorical concepts: we begin the first example by introducing the base language used throughout the paper, *While*, and gradually present distributive laws when required. From the second example in Sect. 6 and on, we relax the categorical notation and mostly remain within the style of PL semantics.

---

[1] The authors use the term "Mathematical Operational Semantics". The term "Bialgebraic Semantics" is also used in the literature.

2 The Basic *While* Language

2.1 Syntax and Operational Semantics

We begin by defining the set of arithmetic expressions.

$$\langle expr \rangle :: = \texttt{lit}\,\mathbb{N}\,|\,\texttt{var}\,\mathbb{N}\,|\,\langle expr \rangle\,\langle bin \rangle\,\langle expr \rangle\,|\,\langle un \rangle\,\langle expr \rangle$$

The constructors are respectively literals, a dereference operator var, binary arithmetic operations as well as unary operations. We let S be the set of lists of natural numbers. The role of S is that of a run-time store whose entries are referred by their index on the list using constructor var. We define function $\texttt{eval} : S \times E \to \mathbb{N}$ inductively on the structure of expressions.

Definition 1 (Evaluation of expressions in *While*)

$$\texttt{eval}\ store\ (\texttt{lit}\ n) = n$$
$$\texttt{eval}\ store\ (\texttt{var}\ l) = \texttt{get}\ store\ l$$
$$\texttt{eval}\ store\ (e_1\ b\ e_2) = (\texttt{eval}\ store\ e_1)\ [[b]]\ (\texttt{eval}\ store\ e_2)$$
$$\texttt{eval}\ store\ (u\ e) = [[u]]\ (\texttt{eval}\ store\ e)$$

Programs in *While* language are generated by the following grammar:

$$\langle prog \rangle :: = \texttt{skip}\,|\,\mathbb{N} := \langle expr \rangle\,|\,\langle prog \rangle\,;\,\langle prog \rangle\,|\,\texttt{while}\,\langle expr \rangle\,\langle prog \rangle$$

The operational semantics of our *While* language are introduced in Fig. 1. We are using the notation $s, x \Downarrow s'$ to denote that program x, when supplied with $s : S$, terminates producing store s'. Similarly, $s, x \to s', x'$ means that program x, supplied with s, evaluates to x' and produces new store s'.

$$\frac{}{s, \texttt{skip} \Downarrow s} \qquad \frac{}{s, l := e \Downarrow \texttt{update}\ s\ l\ (\texttt{eval}\ s\ e)} \qquad \frac{s, p \Downarrow s'}{s, p; q \to s', q}$$

$$\frac{s, p \to s', p'}{s, p; q \to s', p'; q} \qquad \frac{\texttt{eval}\ s\ e = 0}{s, \texttt{while}\ e\ p \to s, \texttt{skip}} \qquad \frac{\texttt{eval}\ s\ e \neq 0}{s, \texttt{while}\ e\ p \to s, p; \texttt{while}\ e\ p}$$

Fig. 1. Semantics of the *While* language.

2.2 *While*, Categorically

The categorical representation of operational semantics has various forms of incremental complexity but for our purposes we only need to use the most important one, that of *GSOS laws* [48].

Definition 2. *Given a syntax functor* Σ *and a behavior functor* B*, a GSOS law of* Σ *over* B *is a natural transformation* $\rho : \Sigma(\mathrm{Id} \times B) \Longrightarrow B\Sigma^*$*, where* (Σ^*, η, μ) *is the monad freely generated by* Σ*.*

Example 1. Let E be the set of expressions of the *While*-language. Then the syntax functor $\Sigma : \mathbf{Set} \to \mathbf{Set}$ for *While* is given by $\Sigma X = \top \uplus (\mathbb{N} \times E) \uplus (X \times X) \uplus (E \times X)$ where $\uplus$ denotes a disjoint (tagged) union. The elements could be denoted as $\mathtt{skip}$, $l := e$, $x_1; x_2$ and $\mathtt{while}\ e\ x$ respectively. The free monad Σ^* satisfies $\Sigma^* X \cong X \uplus \Sigma \Sigma^* X$, i.e. its elements are programs featuring program variables from X. Since *While*-programs run in interaction with a store and can terminate, the behavior functor is $BX = S \to (S \times \mathrm{Maybe}\ X)$, where S is the set of lists of natural numbers and $X \to Y$ denotes the exponential object (internal Hom) Y^X.

The GSOS specification of *While* determines ρ. A premise $s, p \to s', p'$ denotes an element $(p, b) \in (\mathrm{Id} \times B)X$ where $b(s) = (s', \mathrm{just}\ p')$, and a premise $s, p \Downarrow s'$ denotes an element (p, b) where $b(s) = (s', \mathrm{nothing})$. A conclusion $s, p \to s', p'$ (where $p \in \Sigma X$ is further decorated above the line to $\bar{p} \in \Sigma(\mathrm{Id} \times B)X$) specifies that $\rho(\bar{p}) \in B\Sigma^* X$ sends s to $(s', \mathrm{just}\ p')$, whereas a conclusion $s, p \Downarrow s'$ specifies that s is sent to $(s', \mathrm{nothing})$. Concretely, $\rho_X : \Sigma(X \times BX) \to B\Sigma^* X$ is the function (partially from [47]):

$$
\begin{aligned}
\mathtt{skip} &\mapsto \lambda s.(s, \mathrm{nothing}) \\
l := e &\mapsto \lambda s.(\mathtt{update}\ s\ l\ (\mathtt{eval}\ s\ e), \mathrm{nothing}) \\
\mathtt{while}\ e\ (x, f) &\mapsto \lambda s.
\begin{cases}
(s, \mathrm{just}\ (x\ ; \mathtt{while}\ e\ x)) & \text{if } \mathtt{eval}\ s\ e \neq 0 \\
(s, \mathrm{just}\ (\mathtt{skip})) & \text{if } \mathtt{eval}\ s\ e = 0
\end{cases} \\
(x, f)\ ;\ (y, g) &\mapsto \lambda s.
\begin{cases}
(s', \mathrm{just}\ (x'\ ; y)) & \text{if } f(s) = (s', \mathrm{just}\ x') \\
(s', \mathrm{just}\ y) & \text{if } f(s) = (s', \mathrm{nothing})
\end{cases}
\end{aligned}
$$

It has been shown by Lenisa et al. [28] that there is a one-to-one correspondence between GSOS laws of Σ over B and *distributive laws* of the free monad Σ^* over the cofree copointed endofunctor [28] $\mathrm{Id} \times B$.[2]

Definition 3 (In [26]). *A distributive law of a monad (T, η, μ) over a copointed functor (H, ϵ) is a natural transformation $\lambda : TH \Longrightarrow HT$ subject to the following laws: $\lambda \circ \eta = H\eta$, $\epsilon \circ \lambda = T\epsilon$ and $\lambda \circ \mu = H\mu \circ \lambda \circ T\lambda$.*

Given any GSOS law, it is straightforward to obtain the corresponding distributive law via structural induction (In [50], prop. 2.7 and 2.8). By convention, we shall be using the notation ρ for GSOS laws and ρ^* for the equivalent distributive laws unless stated otherwise.

A distributive law λ based on a GSOS law ρ gives a category λ-Bialg of λ-bialgebras [48], which are pairs $\Sigma X \xrightarrow{h} X \xrightarrow{k} BX$ subject to the pentagonal law $k \circ h = Bh^* \circ \rho_X \circ \Sigma[id, k]$, where h^* is the inductive extension of h. Morphisms in λ-Bialg are arrows $X \to Y$ that are both algebra and coalgebra homomorphisms at the same time. The trivial initial B-coalgebra $\bot \to B\bot$ lifts uniquely to the

---

[2] A copointed endofunctor is an endofunctor F equipped with a natural transformation $F \Longrightarrow \mathrm{Id}$.

initial λ-bialgebra $\Sigma\Sigma^*\bot \xrightarrow{a} \Sigma^*\bot \xrightarrow{h_\lambda} B\Sigma^*\bot$, while the trivial final Σ-algebra $\Sigma\top \to \top$ lifts uniquely to the final λ-bialgebra $\Sigma B^\infty\top \xrightarrow{g_\lambda} B^\infty\top \xrightarrow{z} BB^\infty\top^3$. Since $\Sigma^*\bot$ is the set of programs generated by Σ and $B^\infty\top$ the set of behaviors cofreely generated by B, the unique bialgebra morphism $f : \Sigma^*\bot \to B^\infty\top$ is the *interpretation function* induced by ρ.

Remark 1. We write A for $\Sigma^*\bot$ and Z for $B^\infty\top$, and refer to $h_\lambda : A \to BA$ as the *operational model* for λ and to $g_\lambda : \Sigma Z \to Z$ as the *denotational model* [48]. Note also that $a : \Sigma A \cong A$ and $z : Z \cong BZ$ are invertible.

Example 2. Continuing Example 1, the initial bialgebra A is just the set of all *While*-programs. Meanwhile, the final bialgebra Z, which has the meaning of the set of behaviors, satisfies $Z \cong (S \to S \times \text{Maybe } Z)$. In other words, our attacker model is that of an attacker who can count execution steps and moreover, between any two steps, read out and modify the state. In Sect. 9, we discuss how we hope to consider weaker attackers in the future.

3 An Extra Register (Part I)

Let us consider the scenario where a malicious party can observe *more* information about the execution state of a program, either because information is being leaked to the environment or the programs are run by a more powerful machine. A typical example is the presence of an extra *flags* register that logs the result of a computation [8,34,35]. This is the intuition behind the augmented version of *While* with additional observational capabilities, *While*$_\Delta$.

The main difference is in the behavior so the notation for transitions has to slightly change. The two main transition types, $s, \mathbf{x} \Downarrow_v s'$ and $s, x \to_v s', x'$ work similarly to *While* except for the label $v : \mathbb{N}$ produced when evaluating expressions. We also allow language terms to interact with the labels by introducing the constructor **obs** $\mathbb{N}$ $\langle prog \rangle$. When terms evaluate inside an **obs** block, the labels are sequentially placed in the run-time store. The rest of the constructors are identical but the distinction between the two languages should be clear.

While the expressions are the same as before, the syntax functor is now $\Sigma_\Delta X = \Sigma X \uplus \mathbb{N} \times X$, and the behavior functor is $B_\Delta = S \to \mathbb{N} \times S \times \text{Maybe } X$. The full semantics can be found in Fig. 2. As for *While*, they specify a GSOS law $\rho_\Delta : \Sigma_\Delta(\text{Id} \times B_\Delta) \Longrightarrow B_\Delta \Sigma_\Delta^*$.

Traditionally, the (in)security of a compiler has been a matter of *full abstraction*; a compiler is fully abstract if it preserves and reflects Morris-style [31] contextual equivalence. For our threat model, where the attacker can directly observe labels, it makes sense to define contextual equivalence in *While*$_\Delta$ as:

---

[3] We write B^∞ for the cofree comonad over B, which satisfies $B^\infty X \cong X \times BB^\infty X$.

$$\frac{}{s, \text{skip} \Downarrow_0 s} \qquad \frac{v = \text{eval } s\ e}{s, l := e \Downarrow_v \text{update } s\ l\ v} \qquad \frac{v = \text{eval } s\ e \quad v \neq 0}{s, \text{while } e\ p \to_v s, \text{skip}}$$

$$\frac{s, p \Downarrow_v s'}{s, p; q \to_v s', q} \qquad \frac{s, p \Downarrow_v s' \quad s'' = \text{update } s'\ n\ v}{s, \text{obs } n\ p \to_v s'', \text{skip}} \qquad \frac{s, p \to_v s', p'}{s, p; q \to_v s', p'; q}$$

$$\frac{s, p \to_v s', p' \quad s'' = \text{update } s'\ n\ v}{s, \text{obs } n\ p \to_v s'', \text{obs } (n+1)\ p'} \qquad \frac{v = \text{eval } s\ e \quad v = 0}{s, \text{while } e\ p \to_v s, p; \text{while } e\ p}$$

Fig. 2. Semantics of $While_\Delta$.

Definition 4. $p \cong_\Delta q \iff \forall c : C_\Delta.\ c\,[\![p]\!] \Downarrow \iff c\,[\![q]\!] \Downarrow$

Where C is the set of one-hole contexts, $[\![\_]\!] : C_\Delta \times A_\Delta \to A_\Delta$ denotes the plugging function and we write $p \Downarrow$ when p eventually terminates. Contextual equivalence for $While$ is defined analogously. It is easy to show that the simple "embedding" compiler from $While$ to $While_\Delta$ is not fully abstract by examining terms $a \triangleq \text{while } (\text{var}[0])\ (0 := 0)$ and $b \triangleq \text{while } (\text{var}[0] * 2)\ (0 := 0)$, for which $a \cong b$ but $a_\Delta \not\cong_\Delta b_\Delta$. A context $c \triangleq (\text{obs } 1\ \_); \text{while } (\text{var}[1] - 1)\ \text{skip}$ will log the result of the while condition in a_Δ and b_Δ in $\text{var}[1]$ and then either diverge or terminate depending on the value of $\text{var}[1]$. An initial $\text{var}[0]$ value of 1 will cause $c\,[\![a]\!]$ to terminate but $c\,[\![b]\!]$ to diverge.

Securely Extending $While_\Delta$. To deter malicious contexts from exploiting the extra information, we introduce *sandboxing* primitives to help hide it. We add an additional constructor in $While_\Delta$, $\wr\langle progr\rangle\wr$, and the following inference rules to form the secure version $While_\not\Delta$ of $While_\Delta$.

$$\frac{s, p \Downarrow_v s'}{s, \wr p\wr \Downarrow_0 s'} \qquad \frac{s, p \to_v s', p'}{s, \wr p\wr \to_0 s', \wr p'\wr}$$

We now consider the compiler from $While$ to $While_\not\Delta$ which, along with the obvious embedding, wraps the translated terms in sandboxes. This looks to be effective as programs a and b are now contextually equivalent and the extra information is adequately hidden. We will show that this compiler is indeed a *map of distributive laws* between $While$ and $While_\not\Delta$ but to do so we need a brief introduction on the underlying theory.

4   Secure Compilers, Categorically

4.1   Maps of Distributive Laws

Assume two GSOS laws $\rho_1 : \Sigma_1(\text{Id} \times B_1) \implies B_1\Sigma_1^*$ and $\rho_2 : \Sigma_2(\text{Id} \times B_2) \implies B_2\Sigma_2^*$, where $(\Sigma_1^*, \eta_1, \mu_1)$ and $(\Sigma_2^*, \eta_2, \mu_2)$ are the monads freely generated by Σ_1 and Σ_2 respectively. We shall regard pairs of natural transformations $(\sigma : \Sigma_1^* \implies \Sigma_2^*, b : B_1 \implies B_2)$ as compilers between the two semantics, where σ acts as a syntactic translation and b as a translation between behaviors.

Remark 2. If A_1 and A_2 are the sets of terms freely generated by Σ_1 and Σ_2, we can get the compiler $c : A_1 \to A_2$ from σ. On the other hand, b generates a function $d : Z_1 \to Z_2$ between behaviors via finality.

Remark 3. We shall be writing B^c for the cofree copointed endofunctor $\mathrm{Id} \times B$ over B and $b^c : B_1^c \Longrightarrow B_2^c$ for $\mathrm{id} \times b$.

Definition 5 (Adapted from [50]). *A map of GSOS laws from ρ_1 to ρ_2 consists of a natural transformation $\sigma : \Sigma_1^* \Longrightarrow \Sigma_2^*$ subject to the monad laws $\sigma \circ \eta_1 = \eta_2$ and $\sigma \circ \mu_1 = \mu_2 \circ \Sigma_2^* \sigma \circ \sigma$ paired with a natural transformation $b : B_1 \Longrightarrow B_2$ that satisfies the following coherence criterion:*

$$
\begin{array}{ccc}
\Sigma_1^* B_1^c & \overset{\rho_1^*}{\Longrightarrow} & B_1^c \Sigma_1^* \\
{\scriptstyle \sigma \, \circ \, \Sigma_1^* b^c} \Big\Downarrow & & \Big\Downarrow {\scriptstyle b^c \, \circ \, B_1^c \sigma} \\
\Sigma_2^* B_2^c & \overset{\rho_2^*}{\Longrightarrow} & B_2^c \Sigma_2^*
\end{array}
$$

Remark 4. A natural transformation $\sigma : \Sigma_1^* \Longrightarrow \Sigma_2^*$ subject to the monad laws is equivalent to a natural transformation $t : \Sigma_1 \Longrightarrow \Sigma_2^*$.

Theorem 1. *If σ and b constitute a map of GSOS laws, then we get a compiler $c : A_1 \to A_2$ and behavior transformation $d : Z_1 \to Z_2$ satisfying $d \circ f_1 = f_2 \circ c :$ $A_1 \to Z_2$. As bisimilarity is exactly equality in the final coalgebra (i.e. equality under $f_i : A_i \to Z_i$), c preserves bisimilarity [50]. If d is a monomorphism (which, under mild conditions, is the case in* **Set** *if every component of b is a monomorphism), then c also reflects bisimilarity.*

What is very important though, is that the well-behavedness properties of the two GSOS laws bestow *contextual* meaning to bisimilarity. Recall that the gold standard for secure compilation is contextual equivalence (Definition 4), which is precisely what is observable through program contexts. Bisimilarity is generally not the same as contextual equivalence, but we can instead show that in the case of GSOS laws or other forms of distributive laws, bisimilarity defines the *upper bound* (most fine-grained distinction) of observability up to program contexts. We shall do so abstractly in the next subsections.

4.2   Abstract Program Contexts

The informal notion of a *context* in a programming language is that of a program with a hole [31]. Thus contexts are a syntactic construct that models external interactions with a program: a single context is an experiment whose outcome is the evaluation of the subject program *plugged* in the context.

Naïvely, one may hope to represent contexts by a functor H sending a set of variables X to the set HX of terms in ΣX that may have holes in them. A complication is that contexts may have holes at any depth (i.e. any number of operators may have been applied to a hole), whereas ΣX is the set of terms that have exactly one operator in them, immediately applied to variables. One solution is to think of Y in HY as a set of variables that do not stand for terms, but for contexts. This approach is fine for multi-hole contexts, but if we also want to consider single-hole contexts and a given single-hole context c is not the hole itself, then precisely one variable in c should stand for a single-hole context, and all other variables should stand for terms. Thus, in order to support both single- and multi-hole contexts, we make H a two-argument functor, where $H_X Y$ is the set of contexts with term variables from X and context variables from Y.

Definition 6. *Let $\mathbb{C}$ be a distributive category [14] with products $\times$, coproducts $\uplus$, initial object $\perp$ and terminal object $\top$, as is the case for **Set**. A context functor for a syntax functor $\Sigma : \mathbb{C} \to \mathbb{C}$, is a functor $H : \mathbb{C} \times \mathbb{C} \to \mathbb{C}$ (with application to (X, Y) denoted as $H_X Y$) such that there exist natural transformations* hole $:$ $\forall (X, Y).\top \to H_X Y$ *and* con $: \forall X.X \times H_X X \to X \uplus \Sigma X$ *making the following diagram commute for all X:*

$$
\begin{array}{ccc}
X \times \top & \xrightarrow[\cong]{\pi_1} & X \\
{\scriptstyle \mathrm{id}_X \times \mathrm{hole}_{(X,X)}} \downarrow & & \downarrow {\scriptstyle i_1} \\
X \times H_X X & \xrightarrow{\mathrm{con}_X} & X \uplus \Sigma X
\end{array}
$$

The idea of the transformation con is the following: it takes as input a variable $x \in X$ to be plugged into the hole, and a context $c \in H_X X$ with one layer of syntax. The functor H_X is applied again to X rather than Y because x is assumed to have been recursively plugged into the context placeholders $y \in Y$ already. We then make a case distinction: if c is the hole itself, then $i_1\, x$ is returned. Otherwise, $i_2\, c$ is returned.

Definition 7. *Let $\mathbb{C}$ be a category as in Definition 6 and assume a syntax functor Σ with context functor H. If Σ has an initial algebra (A, q_A) (the set of programs) and H_A has a strong initial algebra (C_A, q_{C_A}) [23] (the set of contexts), then we define the plugging function $[\![\ ]\!] : A \times C_A \to A$ as the "strong inductive extension" [23] of the algebra structure $[\mathrm{id}_A, q_A] \circ \mathrm{con}_A : A \times H_A A \to A$ on A, i.e. as the unique morphism that makes the following diagram commute:*

$$
\begin{array}{ccc}
A \times H_A C_A & \xrightarrow[\cong]{id \times q_{C_A}} & A \times C_A \\
{\scriptstyle (\pi, st)} \downarrow & & \\
A \times H_A(A \times C_A) & & \Big\downarrow {\scriptstyle [\![\ ]\!]} \\
{\scriptstyle id \times H_A [\![\ ]\!]} \downarrow & & \\
A \times H_A A & \xrightarrow{\mathrm{con}_A} A \uplus \Sigma A \xrightarrow{[id, q_A]} & A
\end{array}
$$

The above definition of contextual functors is satisfied by both single-hole and multi-hole contexts, the construction of which we discuss below.

Multi-hole Contexts. Given a syntax functor Σ, its multi-hole context functor is simply $H_X Y = \top \uplus \Sigma Y$. The contextual natural transformation con is the obvious map that returns the pluggee if the given context is a hole, and otherwise the context itself (which is then a program):

$$\text{con} : \forall X. X \times (\top \uplus \Sigma X) \to X \uplus \Sigma X$$

$$\text{con} \circ (\text{id} \times i_1) = i_1 \circ \pi_1 : \forall X. X \times \top \to X \uplus \Sigma X$$

$$\text{con} \circ (\text{id} \times i_2) = i_2 \circ \pi_2 : \forall X. X \times \Sigma X \to X \uplus \Sigma X$$

The 'pattern matching' is justified by distributivity of $\mathbb{C}$. For hole $= i_1 : \top \to \top \uplus \Sigma X$, we can see that $\text{con} \circ (\text{id} \times \text{hole}) = i_1 \circ \pi_1$ as required by the definition of a context functor.

Single-Hole Contexts. It was observed by McBride [29] that for inductive types, i.e. least fixpoints/initial algebras μF of certain endofunctors F called *containers* [4] or simply *polynomials*, their single-hole contexts are lists of $\partial F(\mu F)$ where ∂F is the *derivative* of F^4. Derivatives for containers, which were developed by Abbott et al. in [5], enable us to give a categorical interpretation of single-hole contexts as long as the syntax functor Σ is a container.

It would be cumbersome to lay down the entire theory of containers and their derivatives, so we shall instead focus on the more restricted set of *Simple Polynomial Functors* [22] (or SPF), used to model both syntax and behavior. Crucially, SPF's are differentiable and hence compatible with McBride's construction.

Definition 8 (Simple Polynomial Functors). *The collection of SPF is the least set of functors $\mathbb{C} \to \mathbb{C}$ satisfying the following rules:*

$$id \frac{}{\text{Id} \in \text{SPF}} \qquad const \frac{J \in \text{Obj}(\mathbb{C})}{K_J \in \text{SPF}} \qquad prod \frac{F, G \in \text{SPF}}{F \times G \in \text{SPF}}$$

$$coprod \frac{F, G \in \text{SPF}}{F \uplus G \in \text{SPF}} \qquad comp \frac{F, G \in \text{SPF}}{F \circ G \in \text{SPF}}$$

We can now define the differentiation action $\partial : \text{SPF} \to \text{SPF}$ by structural induction. Interestingly, it resembles simple derivatives for polynomial functions.

Definition 9 (SPF derivation rules)

$$\partial \text{Id} = \top, \quad \partial K_J = \bot, \quad \partial(G \uplus H) = \partial G \uplus \partial H,$$

$$\partial(G \times H) = (\partial G \times H) \uplus (G \times \partial H), \quad \partial(G \circ H) = (\partial G \circ H) \times \partial H.$$

---

[4] The *list* operator itself arises from the derivative of the free monad operator.

Example 3. The definition of con for single-hole contexts might look a bit cryptic at first sight so we shall use a small example from [29] to shed some light. In the case of binary trees, locating a hole in a context can be thought of as traversing through a series of nodes, choosing left or right according to the placement of the hole until it is found. At the same time a record of the trees at the non-chosen branches must be kept so that in the end the entire structure can be reproduced.

Now, considering that the set of binary trees is the least fixed point of functor $\top \uplus (\mathrm{Id} \times \mathrm{Id})$, then the type of "abstract" choice at each intersection is the functor $K_{\mathrm{Bool}} \times \mathrm{Id}$, where K_{Bool} stands for a choice of left or right and the Id part represents the passed structure. Lists of $(K_{\mathrm{Bool}} \times \mathrm{Id})$ BinTree are exactly the sort of record we need to keep, i.e. they contain the same information as a tree with a single hole. And indeed $K_{\mathrm{Bool}} \times \mathrm{Id}$ is (up to natural isomorphism) the derivative of $\top \uplus (\mathrm{Id} \times \mathrm{Id})$!

Using derivatives we can define the context functor $H_X Y = \top \uplus ((\partial \Sigma\, X) \times Y)$ for syntax functor Σ. Then the initial algebra C_A of H_A is indeed List $((\partial \Sigma)\, A)$, the set of single-hole contexts for $A \cong \Sigma A$.

Plugging. Before defining con, we define an auxiliary function conStep : $\partial \Sigma \times \mathrm{Id} \implies \Sigma$. We defer the reader to [29] for the full definition of conStep, which is inductive on the SPF Σ, and shall instead only define the case for coproducts. So, for $\partial(F \uplus G) = \partial F \uplus \partial G$ we have:

$$\mathrm{conStep}_{F \uplus G} : (\partial F \uplus \partial G) \times \mathrm{Id} \implies F \uplus G$$
$$\mathrm{conStep}_{F \uplus G} \circ (i_1 \times \mathrm{id}) = i_1 \circ \mathrm{conStep}_F : \partial F \times \mathrm{Id} \implies F \uplus G$$
$$\mathrm{conStep}_{F \uplus G} \circ (i_2 \times \mathrm{id}) = i_2 \circ \mathrm{conStep}_G : \partial G \times \mathrm{Id} \implies F \uplus G$$

We may now define con : $X \times H_X X \to X \uplus \Sigma X$ as follows:

$$\mathrm{con} : \forall X.X \times (\top \uplus (\partial \Sigma\, X \times X)) \to X \uplus \Sigma X$$
$$\mathrm{con} \circ (\mathrm{id} \times i_1) = i_1 \circ \pi_1 : \forall X.X \times \top \to X \uplus \Sigma X$$
$$\mathrm{con} \circ (\mathrm{id} \times i_2) = i_2 \circ \mathrm{conStep}_\Sigma \circ \pi_2 : \forall X.X \times (\partial \Sigma\, X \times X) \to X \uplus \Sigma X$$

By setting hole $= i_1 : \top \to \top \uplus (\partial \Sigma\, X \times X)$ we can see that $\mathrm{con} \circ (\mathrm{id} \times \mathrm{hole}) = i_1 \circ \pi_1$ as required by Definition 6.

4.3   Contextual Coclosure

Having established a categorical notion of contexts, we can now move towards formulating contextual categorical arguments about bisimilarity. We assume a context functor H for Σ such that H_A has strong initial algebra (C_A, q_{C_A}) (the object containing all contexts).

First, since we prefer to work in more general categories than just **Set**, we will encode relations $R \subseteq X \times Y$ as spans $X \xleftarrow{r_1} R \xrightarrow{r_2} Y$. One may wish to consider only spans for which $(r_1, r_2) : R \to X \times Y$ is a monomorphism, though this is not necessary for our purposes.

We want to reason about contextually closed relations on the set of terms A, which are relations such that $a_1 \mathrel{R} a_2$ implies $(c\,\llbracket a_1 \rrbracket) \mathrel{R} (c\,\llbracket a_2 \rrbracket)$ for all contexts $c \in C_A$. Contextual equivalence will typically be defined as the co-closure of equitermination: the greatest contextually closed relation that implies equitermination. For spans, this becomes:

Definition 10. *In a category as in Definition 6, a span $A \xleftarrow{r_1} R \xrightarrow{r_2} A$ is called contextually closed if there is a morphism $\llbracket\,\rrbracket : C_A \times R \to R$ making the following diagram commute:*

$$
\begin{array}{ccccc}
C_A \times A & \xleftarrow{\mathrm{id} \times r_1} & C_A \times R & \xrightarrow{\mathrm{id} \times r_2} & C_A \times A \\
\downarrow{\scriptstyle \llbracket\,\rrbracket} & & \downarrow{\scriptstyle \llbracket\,\rrbracket} & & \downarrow{\scriptstyle \llbracket\,\rrbracket} \\
A & \xleftarrow{\quad r_1 \quad} & R & \xrightarrow{\quad r_2 \quad} & A
\end{array}
$$

The contextual co-closure $A \xleftarrow{\bar{r}_1} \bar{R} \xrightarrow{\bar{r}_2} A$ of an arbitrary span $A \xleftarrow{r_1} R \xrightarrow{r_2} A$ is the final contextually closed span on A with a span morphism $\bar{R} \to R$. ⌟

We call terms bisimilar if the operational semantics $f : A \to Z$ assigns them equal behaviors:

Definition 11. *We define (strong) bisimilarity $\sim_{\text{bis}}$ as the pullback of the equality span $(\mathrm{id}_Z, \mathrm{id}_Z) : Z \to Z \times Z$ along $f \times f : A \times A \to Z \times Z$ (if existent).*

Theorem 2. *Under the assumptions of Definition 7, bisimilarity (if existent) is contextually closed.*

Proof. We need to give a morphism of spans from $C_A \times (\sim_{\text{bis}})$ to $(\sim_{\text{bis}})$:

$$
\begin{array}{ccccc}
C_A \times A & \xleftarrow{\mathrm{id} \times r_1} & C_A \times (\sim_{\text{bis}}) & \xrightarrow{\mathrm{id} \times r_2} & C_A \times A \\
\downarrow{\scriptstyle \llbracket\,\rrbracket} & & \vdots & & \downarrow{\scriptstyle \llbracket\,\rrbracket} \\
A & \xleftarrow{\quad r_1 \quad} & (\sim_{\text{bis}}) & \xrightarrow{\quad r_2 \quad} & A \\
\downarrow{\scriptstyle f} & & \downarrow{\scriptstyle w} & & \downarrow{\scriptstyle f} \\
Z & \xleftarrow{\quad \mathrm{id}_Z \quad} & Z & \xrightarrow{\quad \mathrm{id}_Z \quad} & Z.
\end{array}
$$

By definition of $(\sim_{\text{bis}})$, it suffices to give a morphism of spans to the equality span on Z, i.e. to prove that $f \circ \llbracket\,\rrbracket \circ (\mathrm{id} \times r_1) = f \circ \llbracket\,\rrbracket \circ (\mathrm{id} \times r_2)$. To this end, consider the following diagram (parameterized by $i \in \{1,2\}$), in which every polygon is easily seen to commute:

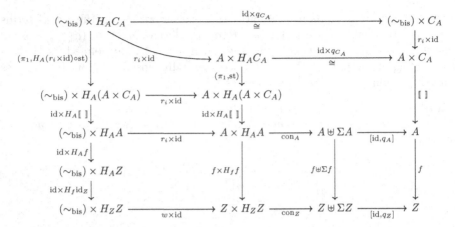

The bottom-right square stems from the underlying GSOS law: it is the algebra homomorphism part of the bialgebra morphism between the initial and the final bialgebras. Commutativity of the outer diagram reveals that $f \circ [\![\ ]\!] \circ (r_i \times \mathrm{id})$ is, regardless of i, *the* strong inductive extension of $[\mathrm{id}_Z, q_Z] \circ \mathrm{con}_Z \circ (w \times H_f \mathrm{id}_Z) :$ $(\sim_{\mathrm{bis}}) \times H_A Z \to Z$. Thus, it is independent of i.   □

Corollary 1. *In* **Set**, *bisimilarity is its own contextual coclosure:* $a \sim_{\mathrm{bis}} b \iff$ $\forall\ c \in C_A.\ c[\![a]\!] \sim_{\mathrm{bis}} c[\![b]\!]$.

Corollary 2. *In* **Set**, *bisimilarity implies contextual equivalence.*[5]

Proof. Bisimilarity implies equitermination. This yields an implication between their coclosures.   □

Comparing Corollary 1 to contextual equivalence in Definition 4 reveals their key difference. Contextual equivalence makes minimal assumptions on the underlying observables, which are simply divergence and termination. On the other hand, the contextual coclosure of bisimilarity assumes maximum observability (as dictated by the behavior functor) and in that sense it represents the upper bound of what can be observed through contexts. Consequently, this criterion is useful if the observables adequately capture the threat model, which is true for the examples that follow.

This theorem echoes similar results in the broader study of coalgebraic bisimulation [11,41]. There are, however, two differences. The first is that our theorem allows for extra flexibility in the definition of contexts as the theorem is parametric on the context functor. Second, by making the context construction explicit we can directly connect (the contextual coclosure of) bisimilarity to contextual equivalence (Corollary 2) and so have a more convincing argument for using maps of distributive laws as secure compilers.

---

[5] Note that we can *not* conclude that preservation of bisimilarity would imply preservation of contextual equivalence.

5   An Extra Register (Part II)

The next step is to define the syntax and behavior natural transformations. The first compiler, $\sigma_\triangle : \Sigma \Longrightarrow \Sigma_\triangle$, is a very simple mapping of constructors in *While* to their *While*$_\triangle$ counterparts. The second natural transformation, $\sigma_\boxtimes : \Sigma \Longrightarrow \Sigma^*_\boxtimes$, is more complex as it involves an additional layer of syntax in *While*$_\boxtimes$.

Definition 12 (Sandboxing natural transformation). *Consider the natural transformation* $e : \Sigma \Longrightarrow \Sigma_\boxtimes$ *which embeds* Σ *in* $\Sigma_\boxtimes$. *Using PL notation, we define* $\sigma_\boxtimes : \Sigma X \to \Sigma^*_\boxtimes X : p \mapsto \lceil e(p) \rfloor$. *This yields a monad morphism* $\sigma^*_\boxtimes : \Sigma^* \to \Sigma^*_\boxtimes$ *(Remark 4).*

Defining the natural translation between behaviors is a matter of choosing a designated value for the added observable label. The only constraint is that the chosen value has to coincide with the label that the sandbox produces. $B_\triangle$ and $B_\boxtimes$ are identical so we need a single natural transformation $b : B \Longrightarrow B_{\triangle/\boxtimes}$:

$$b : \forall X. \ (S \to S \times \text{Maybe } X) \to S \to \mathbb{N} \times S \times \text{Maybe } X$$
$$b \ f = \lambda(s : S) \to (0, f(s))$$

While to While$_\triangle$. We now have the natural translation pairs $(\sigma_\triangle, b)$ and $(\sigma_\boxtimes, b)$, which allows us to check the coherence criterion from Sect. 4.1. We shall be using a graphical notation that provides for a good intuition as to what failure or success of the criterion *means*. For example, Fig. 3 shows failure of the coherence criterion for the first pair.

The horizontal arrows in the diagram represent the two semantics, ρ^* and $\rho^*_\triangle$, while the vertical arrows are the two horizontal compositions of the natural translation pair. The top-left node holds an element of $\Sigma^*(\text{Id} \times B)$, which in this case is an assignment operation. The two rightmost nodes represent behaviors, so the syntactic element is missing from the left side of the transition arrows.

$$
\begin{array}{ccc}
l := e & \xrightarrow{\ \rho^*\ } & \dfrac{v = \mathbf{eval}\ s\ e}{s \Downarrow \mathbf{update}\ s\ l\ v} \\[2ex]
\Big\downarrow{\scriptstyle \sigma^*_\triangle \circ \Sigma^* bc} & & \Big\downarrow{\scriptstyle b^c \circ B^c \sigma^*_\triangle} \\[2ex]
l := e & \xrightarrow{\ \rho^*_\triangle\ } & \dfrac{v = \mathbf{eval}\ s\ e}{s \Downarrow_{v/0} \mathbf{update}\ s\ l\ v}
\end{array}
$$

Fig. 3. Failure of the criterion for $(\sigma_\triangle, b)$.

In the upper path, the term is first applied to the GSOS law ρ^* and the result is then passed to the translation pair, thus producing the designated label 0, typeset in blue for convenience. In the lower path, the term is first applied to the translation and then goes through the target semantics, $\rho^*_\triangle$, where the label v is produced. It is easy to find such an s so that $v \neq 0$.

While to *While*$_{\not{\alpha}}$. The same example is investigated for the second translation pair $(\sigma_{\not{\alpha}}, b)$. Figure 4 shows what happens when we test the same case as before. Applying $\rho^*_{\not{\alpha}}$ to $\lfloor l := e \rfloor$ is similar to $\rho_{\not{\alpha}}$ acting twice. The innermost transition is the intermediate step and as it only appears in the bottom path it is

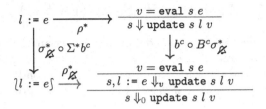

Fig. 4. The coherence criterion for $(\sigma_{\not{\alpha}}, b)$.

typeset in red. This time the diagram commutes as the label produced in the inner layer, v, is effectively erased by the sandboxing rules of *While*$_{\not{\alpha}}$.

An endo-compiler for *While*$_{\not{\alpha}}$. If $A_{\not{\alpha}}$ is the set of closed terms for *While*$_{\not{\alpha}}$, the compiler $u : A_{\not{\alpha}} \to A_{\not{\alpha}}$, which "escapes" *While*$_{\not{\alpha}}$ terms from their sandboxes can be elegantly modeled using category theory. As before, it is not possible to express it using a simple natural transformation $\Sigma_{\not{\alpha}} \implies \Sigma_{\not{\alpha}}$. We can, however, use the *free pointed endofunctor* [28] over $\Sigma_{\not{\alpha}}$, Id $\uplus \Sigma_{\not{\alpha}}$. What we want is to map non-sandboxed terms to themselves and lift the extra layer of syntax from sandboxed terms. Intuitively, for a set of variables X, $\Sigma_{\not{\alpha}}X$ is one layer of syntax "populated" with elements of X. If $X \uplus \Sigma_{\not{\alpha}}X$ is the union of $\Sigma_{\not{\alpha}}X$ with the set of variables X, lifting the sandboxing layer is mapping the X in $\lfloor X \rfloor$ to the left of $X \uplus \Sigma_{\not{\alpha}}X$ and the rest to themselves at the right.

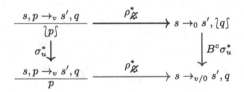

Fig. 5. Failure of the criterion for σ_u.

This is obviously not a secure compiler as it allows discerning previously indistinguishable programs. As we can see in Fig. 5, the coherence criterion fails in the expected manner.

6 State Mismatch

Having established our categorical foundations, we shall henceforth focus on examples. The first one involves a compiler where the target machine is not necessarily more powerful than the source machine, but the target *value* primitives are not isomorphic to the ones used in the source. This is a well-documented problem [37], which has led to failure of full abstraction before [8,18,25].

For example, we can repeat the development of *While* except we substitute natural numbers with integers. We call this new version *While*$_{\mathbb{Z}}$.

$$\langle expr \rangle :: = \texttt{lit}\,\mathbb{Z} \mid \texttt{var}\,\mathbb{N} \mid \langle expr \rangle \, \langle bin \rangle \, \langle expr \rangle \mid \langle un \rangle \, \langle expr \rangle$$

The behavior functor also differs in that the store type S is substituted with $S_\mathbb{Z}$, the set of lists of integers. We can define the behavioral natural transformation $b_\mathbb{Z} : B \Longrightarrow B_\mathbb{Z}$ as the best "approximation" between the two behaviors. In **Set**:

$$b_\mathbb{Z} : \forall X.\ (S \to S \times (\top \uplus X)) \to S_\mathbb{Z} \to S_\mathbb{Z} \times (\top \uplus X)$$

$$b_\mathbb{Z}\, f = [\text{to}\mathbb{Z}, \text{id}] \circ f \circ \text{to}\mathbb{N}$$

Where to$\mathbb{N}$ replaces all negative numbers in the store with 0 and to$\mathbb{Z}$ typecasts S to $S_\mathbb{Z}$. It is easy to see that the identity compiler from *While* to *While$_\mathbb{Z}$* is not fully abstract. For example, the expressions 0 and min(var[0], 0) are identical in *While* but can be distinguished in *While$_\mathbb{Z}$* (if var[0] is negative). This is reflected in the coherence criterion diagram for the identity compiler in Fig. 6, when initiating the store with a negative integer.

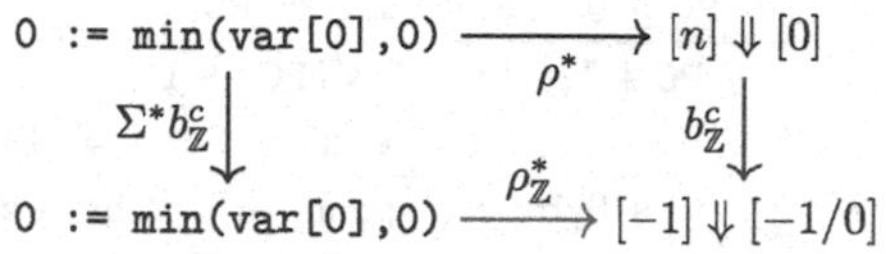

Fig. 6. Failure of the criterion for $(\text{id}, b_\mathbb{Z})$.

The solution is to create a special environment where *While$_\mathbb{Z}$* forgets about negative integers, in essence copying what $b_\mathbb{Z}$ does on the store. This is a special kind of sandbox, written $\langle \_ \rangle$, for which we introduce the following rules:

$$\frac{\text{to}\mathbb{N}(s), p \Downarrow s'}{s, \langle p \rangle \Downarrow s'} \qquad \frac{\text{to}\mathbb{N}(s), p \to s', p'}{s, \langle p \rangle \to s', \langle p' \rangle}$$

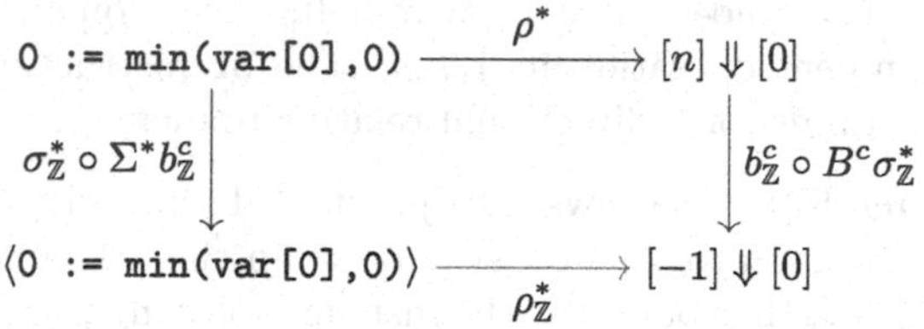

Fig. 7. The coherence criterion for $(\sigma_\mathbb{Z}, b_\mathbb{Z})$.

We may now repeat the construction from Definition 12 to define the compiler $\sigma_\mathbb{Z}$. We can easily verify that the pair $(\sigma_\mathbb{Z}, b_\mathbb{Z})$ constitutes a map of distributive laws. For instance, Fig. 7 demonstrates how the previous failing case now works under $(\sigma_\mathbb{Z}, b_\mathbb{Z})$.

7 Control Flow

Many low-level languages support unrestricted control flow in the form of jumping or branching to an address. On the other hand, control flow in high-level languages is usually restricted (think if-statements or function calls). A compiler from the high-level to the low-level might be insecure as it exposes source-level programs to illicit control flow. This is another important and well-documented example of failure of full abstraction [3,8,33,37].

$$\frac{}{s, 0, \mathsf{stop} \; [; ; \; x] \Downarrow s, 0} \qquad \frac{v = \mathsf{eval} \; s \; e \quad s' = \mathsf{update} \; s \; n \; v}{s, 0, \mathsf{assign} \; n \; v \; [; ; \; x] \to s', 1, \mathsf{assign} \; n \; v \; [; ; \; x]}$$

$$\frac{\mathrm{PC} \geq 0 \quad s, \mathrm{PC}, x \Downarrow s', \mathrm{PC}'}{s, \mathrm{PC} + 1, i \; ; ; \; x \Downarrow s', \mathrm{PC}' + 1} \qquad \frac{v = \mathsf{eval} \; s \; e \quad v = 0}{s, 0, \mathsf{br} \; e \; z \; [; ; \; x] \to s, 1, \mathsf{br} \; e \; z \; [; ; \; x]}$$

$$\frac{v = \mathsf{eval} \; s \; e \quad v \neq 0}{s, 0, \mathsf{br} \; e \; z \; [; ; \; x] \to s, z, \mathsf{br} \; e \; z \; [; ; \; x]} \qquad \frac{\mathrm{PC} < 0}{s, \mathrm{PC}, i \; ; ; \; x \Downarrow s, \mathrm{PC}}$$

$$\frac{p = \mathsf{nop} \; [; ; \; x]}{s, 0, p \to s, 1, p} \qquad \frac{\mathrm{PC} \geq 0 \quad s, \mathrm{PC}, x \to s', \mathrm{PC}', x'}{s, \mathrm{PC} + 1, i \; ; ; \; x \to s', \mathrm{PC}' + 1, i \; ; ; \; x'} \qquad \frac{\mathrm{PC} \neq 0}{s, \mathrm{PC}, i \Downarrow s, \mathrm{PC}}$$

Fig. 8. Semantics of the *Low* language. Elements in square brackets are optional.

We introduce low-level language *Low*, the programs of which are non-empty lists of instructions. *Low* differs significantly from *While* and its derivatives in both syntax and semantics. For the syntax, we define the set of instructions ⟨*inst*⟩ and set of programs ⟨*asm*⟩.

⟨*inst*⟩ :: = $\mathsf{nop} \mid \mathsf{stop} \mid \mathsf{assign} \; \mathbb{N} \; \langle expr \rangle \mid \mathsf{br} \; \langle expr \rangle \; \mathbb{Z}$

⟨*asm*⟩ :: = $\langle inst \rangle \mid \langle inst \rangle \; ; ; \; \langle asm \rangle$

Instruction nop is the no-operation, stop halts execution and assign is analogous to the assignment operation in *While*. The br instruction is what really defines *Low*, as it stands for bidirectional relative branching.

Semantics of Low. Figure 8 shows the operational semantics of *Low*. The execution state of a running program consists of a run-time store and the program counter register $\mathrm{PC} \in \mathbb{Z}$ that points at the instruction being processed. If the program counter is zero, the leftmost instruction is executed. If the program counter is greater than zero, then the current instruction is further to the right. Otherwise, the program counter is out-of-bounds and execution stops. The categorical interpretation suggests a GSOS law ρ_L of syntax functor $\Sigma_L X = \mathsf{inst} \uplus (\mathsf{inst} \times X)$ over behavior functor $B_L X = S \times \mathbb{Z} \to S \times \mathbb{Z} \times \mathsf{Maybe} \; X$.

An Insecure Compiler. This time we start with the behavioral translation, which is less obvious as we have to go from $BX = S \to S \times \mathsf{Maybe} \; X$ to $B_L X = S \times \mathbb{Z} \to S \times \mathbb{Z} \times \mathsf{Maybe} \; X$. The increased arity in B_L poses an interesting question as to what the program counter should mean in *While*. It makes sense to consider the program counter in *While* as zero since a program in *While* is treated uniformly as a single statement.

$$b_L : \forall X.\ (S \to S \times \text{Maybe } X) \to S \times \mathbb{Z} \to S \times \mathbb{Z} \times \text{Maybe } X$$

$$b_L\ f\ (s,0) = \begin{cases} (s',1,\text{nothing}) & \text{if } f\ s = (s',\text{nothing}) \\ (s',0,\text{just } y) & \text{if } f\ s = (s',\text{just } y) \end{cases}$$

$$b_L\ f\ (s,n \neq 0) = (s,n,\text{nothing})$$

When it comes to translating terms, a typical compiler from *While* to *Low* would untangle the tree-like structure of *While* and convert it to a list of *Low* instructions. For `while` statements, the compiler would use branching to simulate looping in the low-level.

Example 4. Let us look at a simple case of a loop. The *While* program
`while (var 0 < 2) (1 := var 1 + 1)` is compiled to
`br !(var 0 < 2) 3 ;; assign 1 (var 1 + 1) ;; br (lit 1) -2`

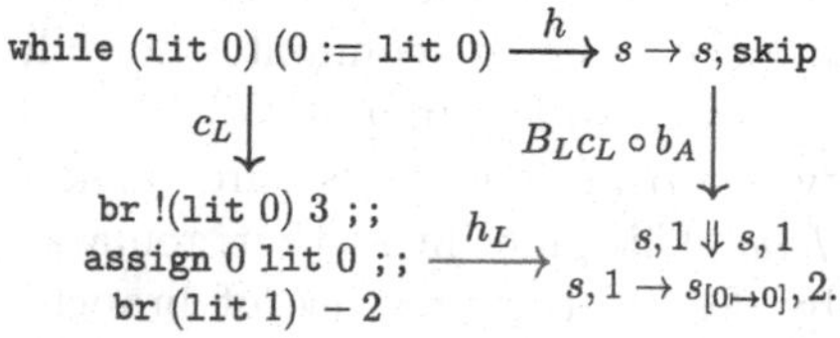

This compiler, called c_L, cannot be defined in terms of a natural transformation $\Sigma \implies \Sigma_L^*$ as per Remark 4, but it is inductive on the terms of the source language. In this case we can directly compare the two operational models $b_A \circ h : A \to B_L A$ (where $h : A \to BA$) and $h_L : A_L \to B_L A_L$ and notice that $c_L : A \to A_L$ is not

Fig. 9. c_L is not a coalgebra homomorphism.

a coalgebra homomorphism (Fig. 9). The key is that the program counter in *Low* allows for finer observations on programs. Take for example the case for `while (lit 0) (0 := lit 0)`, where the loop is always skipped. In *Low*, we can still access the loop body by simply pointing the program counter to it. This is a realistic attack scenario because *Low* allows manipulation of the program counter via the `br` instruction.

Solution. By comparing the semantics between *While* in Fig. 1 and *Low* in Fig. 8 we find major differences. The first one is the reliance of *Low* to a program counter which keeps track of execution, whereas *While* executes statements from left to right. Second, the sequencing rule in *While* dictates that statements are removed from the program state[6] upon completion. On the other hand, *Low* keeps the program state intact at all times. Finally, there is a stark contrast between the two languages in the way they handle `while` loops.

To address the above issues we introduce a new sequencing primitive $;;_c$ and a new looping primitive `loop` for *Low*, which prohibit illicit control flow and properly propagate the internal state. Furthermore, we change the semantics of the singleton `assign` instruction so that it mirrors the peculiarity of its *While* counterpart. The additions can be found in Fig. 10.

---

[6] We are not referring to the store, but to the internal, algebraic state.

$$\frac{\text{PC} \neq 0}{s, \text{PC}, x \mathbin{;;_c} y \Downarrow s', \text{PC}} \qquad \frac{s, 0, x \Downarrow s', z}{s, 0, x \mathbin{;;_c} y \to s', 0, y} \qquad \frac{s, 0, x \to s', z, x'}{s, 0, x \mathbin{;;_c} y \to s', 0, x' \mathbin{;;_c} y}$$

$$\frac{\text{PC} \neq 0}{s, \text{PC}, \text{loop } e\ x \Downarrow s, \text{PC}} \qquad \frac{v = \text{eval } s\ e \quad v = 0}{s, 0, \text{loop } e\ x \to s, 0, \text{stop}}$$

$$\frac{v = \text{eval } s\ e \quad v \neq 0}{s, 0, \text{loop } e\ x \to s, 0, x \mathbin{;;_c} \text{loop } e\ x} \qquad \frac{v = \text{eval } s\ e \quad s' = \text{update } s\ n\ v}{s, 0, \text{assign } n\ v \Downarrow s', 0}$$

Fig. 10. Secure primitives for the *Low* language.

We may now define the simple "embedding" natural transformation $\sigma_E : \Sigma \Longrightarrow \Sigma_L$, which maps skip to stop, assignments to assign, sequencing to $\mathbin{;;_c}$ and while to loop.

Figure 11 shows success of the coherence criterion for the while case. Since the diagram commutes for all cases, (σ_E, b_E) is a map of GSOS laws between *While* and the secure version of *Low*. This guarantees that, remarkably, despite the presence of branching, a low-level attacker cannot illicitly access code that is unreachable on

Fig. 11. The coherence criterion for (σ_E, b_L).

the high-level. Regardless, the solution is a bit contrived in that the new *Low* primitives essentially copy what *While* does. This is partly because the above are complex issues involving radically different languages but also due to the current limitations of the underlying theory. We elaborate further on said limitations, as well as advantages and future improvements, at Sect. 9.

8   Local State Encapsulation

High-level programming language abstractions often involve some sort of private state space that is protected from other objects. Basic examples include functions with local variables and objects with private members. Low-level languages do not offer such abstractions but when it comes to *secure architectures*, there is some type of *hardware sandboxing*[7] to facilitate the need for *local state encapsulation*. Compilation schemes that respect confidentiality properties have been a central subject in secure compilation work [8,18,37,46], dating all the way back to Abadi's seminal paper [1].

In this example we will explore how local state encapsulation fails due to lack of stack clearing [44,46]. We begin by extending *While* to support blocks which have their own private state, thus introducing *While$_B$*. More precisely, we add the frame and return commands that denote the beginning and end of a new block. We also have to modify the original behavior functor B to act on a stack of stores by simply specifying $B_B X = [S] \to [S] \times \text{Maybe } X$, where $[S]$

---

[7] Examples of this are enclaves in Intel SGX [15] and object capabilities in CHERI [51].

denotes a list of stores. For reasons that will become apparent later on, we shall henceforth consider stores of a certain length, say L.

$$\frac{}{m, \texttt{skip} \Downarrow m} \qquad \frac{v = \texttt{eval'} \ m \ e \quad m' = \texttt{update'} \ m \ l \ v}{m, l \ := e \Downarrow m'} \qquad \frac{m, p \Downarrow m'}{m, p; q \to m', q}$$

$$\frac{m, p \to m', p'}{m, p; q \to m', p'; q} \qquad \frac{\texttt{eval'} \ m \ e = 0}{m, \texttt{while} \ e \ p \to m, \texttt{skip}}$$

$$\frac{\texttt{eval'} \ m \ e \neq 0}{m, \texttt{while} \ e \ p \to m, p; \texttt{while} \ e \ p} \qquad \frac{s_0 = [0, 0, \ldots, 0]}{m, \texttt{frame} \Downarrow s_0 :: m} \qquad \frac{}{s :: m, \texttt{return} \Downarrow m}$$

Fig. 12. Semantics of the $While_B$ language.

The semantics for $While_B$ can be found in Fig. 12. Command $\texttt{frame}$ allocates a new private store by appending one to the stack of stores while $\texttt{return}$ pops the top frame from the stack. This built-in, automatic (de)allocation of frames guarantees that there are no traces of activity, in the form of stored values, of past blocks. The rest of the semantics are similar to $While$, only now evaluating an expression and updating the state acts on a stack of stores instead of a single, infinite store and $\texttt{var}$ expressions act on the active, topmost frame.

$$\frac{m' = \texttt{update} \ m \ (l + L * sp) \ (\texttt{evalSP} \ m \ sp \ e)}{(m, sp), l \ := e \Downarrow (m', sp)} \qquad \frac{sp > 0}{(m, sp), \texttt{return} \Downarrow (m, sp - 1)}$$

$$\frac{(m, sp), p \Downarrow (m', sp)}{(m, sp), p; q \to (m', sp), q} \qquad \frac{(m, sp), p \to (m', sp), p'}{(m, sp), p; q \to (m', sp), p'; q}$$

$$\frac{}{(m, sp), \texttt{skip} \Downarrow (m, sp)} \qquad \frac{\texttt{evalSP} \ m \ sp \ e = 0}{(m, sp), \texttt{while} \ e \ p \to (m, sp), \texttt{skip}}$$

$$\frac{\texttt{evalSP} \ m \ sp \ e \neq 0}{(m, sp), \texttt{while} \ e \ p \to (m, sp), p; \texttt{while} \ e \ p} \qquad \frac{}{(m, sp), \texttt{frame} \Downarrow (m, sp + 1)}$$

Fig. 13. Semantics of the $Stack$ language.

Low-Level Stack. In typical low-level instruction sets like the Intel x86 [21] or MIPS [30] there is a single, continuous *memory* which is partitioned in *frames* via processor registers. Figure 13 shows the semantics of $Stack$, a variant of $While_B$ with the same syntax that incorporates a simple *low-level* stack. The difference is that the stack frames are all sized L, the same size as each individual store in $While_B$, so at each $\texttt{frame}$ and $\texttt{return}$ we need only increment and decrement the *stack pointer*. The presence of the stack pointer, which is essentially a natural number, means that the behavior of $Stack$ is $B_S X = S \times \mathbb{N} \to S \times \mathbb{N} \times \text{Maybe} \ X$. The new evaluation function, $\texttt{evalSP}$, works similarly to $\texttt{eval}$ in Definition 1, except for $\texttt{var} \ l$ expressions that dereference values at offset $l + L * sp$.

An Insecure Compiler. $While_B$ and $Stack$ share the same syntax so we only need a behavioral translation, which is all about relating the two different notions of stack. We thus define natural transformation $b_B : B_B \Longrightarrow B_S$:

$$b_B : \forall X.\ ([S_L] \to [S_L] \times \mathrm{Maybe}\ X) \to S \to \mathbb{N} \to S \times \mathbb{N} \times \mathrm{Maybe}\ X$$
$$b_B\ f\ s\ sp = (\mathrm{override}\ (\mathrm{join}\ m)\ s, \mathrm{len}\ m, y)\ \textsf{where}\ (m, y) = f\ (\mathrm{div}\ s\ sp)$$
$$\mathrm{div}\ s\ sp = (\mathrm{take}\ L\ s) :: (\mathrm{div}\ (\mathrm{drop}\ L\ s)\ (sp - 1))$$
$$\mathrm{override}\ s'\ s = s'\ \texttt{++}\ \mathrm{drop}\ (\mathrm{len}\ s')\ s$$

We "divide" an infinite list by the number of stack frames, feed the result to the behavior function f and join ("flatten") it back together while keeping the original part of the infinite list which extends beyond the *active stack* intact. Note that in the case of the `frame` command f adds a new frame to the list of stores. The problem is that in $While_B$ the new frame is initialized to 0 in contrast to $Stack$ where `frame` does not initialize new frames. This leads to a failure of the coherence criterion for (id, b_B) as we can see in Fig. 14.

Failure of the criterion is meaningful in that it underlines key problems of this compiler which can be exploited by a low-level attacker. First, the low-level calling convention indirectly allows terms to access expired stack frames. Second, violating the assumption in $While_B$ that new frames are properly initialized breaks behavioral equivalence. For example, programs $a \triangleq$ `frame ; 0 := var[0]+1` and $b \triangleq$ `frame ; 0 := 1` behave identically in $While_B$ but not in $Stack$.

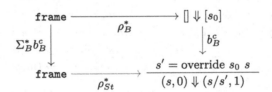

Fig. 14. Failure of the criterion for (id, b_B).

Solution. It is clear that the lack of stack frame initialization in $Stack$ is the lead cause of failure so we introduce the following fix in the `frame` rule.

$$\frac{m' = (\mathrm{take}\ (L * sp)\ m)\ \texttt{++}\ s_0\ \texttt{++}\ (\mathrm{drop}\ ((L + 1)^` * sp)\ m)}{(m, sp), \textsf{frame} \Downarrow (m', sp + 1)}$$

The idea behind the new `frame` rule is that the L-sized block in position sp, which is going to be the new stack frame, has all its values replaced by zeroes. As we can see in Fig. 15, the coherence criterion is now satisfied and the example described earlier no longer works.

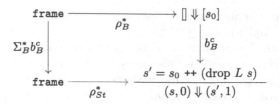

Fig. 15. The coherence criterion for (id, b_B) under the new `frame` rule.

9   Discussion and Future Work

On Mathematical Operational Semantics. The cases we covered in this paper are presented using Plotkin's Structural Operational Semantics [38], yet their foundations are deeply categorical [48]. Consequently, for one to use the methods presented in this paper, the semantics involved must fall within the framework of distributive laws, the generality of which has been explored in the past [47,50], albeit not exhaustively. To the best of our knowledge, Sect. 7 and Sect. 8 show the first instances of distributive laws as low-level machines.

Bialgebraic semantics are well-behaved in that *bisimilarity* is a *congruence* [19]. We used that to show that two bisimilar programs will remain bisimilar irrespective of the context they are plugged into, which is not the same as contextual equivalence. However, full abstraction is but one of a set of proposed characterizations of secure compilation [2,36] and the key intuition is that our framework is suitable as long as bisimilarity adequately captures the threat model. While this is the case in the examples, we can imagine situations where the threat model is *weaker* than the one implied by bisimilarity.

For example, language *While*$_\Omega$ in Sect. 3 includes labels in its transition structure and the underlying model is accurate in that *While*$_\Omega$ terms can manipulate said labels. However, if we were to remove **obs** statements from the syntax, the threat model becomes weaker than the one implied by bisimilarity. Similarly in Sect. 7 and *Low*, we could remove the implicit assumption that the program counter can be manipulated by a low-level attacker.

This issue can be classified as part of the broader effort towards coalgebraic weak bisimilarity, a hard problem which has been an object of intense, ongoing scientific research [12,13,20,39,42,43]. Of particular interest is the work by Abou-Saleh and Pattinson [6,7] about bialgebraic semantics, where they use techniques introduced in [20] to obtain a more appropriate semantic domain for effectful languages as a final coalgebra in the Kleisli category of a suitable monad. This method is thus a promising avenue towards exploring weaker equivalences in bialgebraic semantics, as long as these can be described by a monad.

On Maps of Distributive Laws. Maps of distributive laws were first mentioned by Power and Watanabe [40], then elaborated as *Well-behaved translations* by Watanabe [50] and more recently by Klin and Nachyla [27]. Despite the few examples presented in [27,50], this paper is the first major attempt towards applying the theory behind maps of distributive laws in a concrete problem, let alone in secure compilation.

From a theoretical standpoint, maps of distributive laws have remained largely the same since their introduction. This comes despite the interesting developments discussed in Sect. 9 regarding distributive laws, which of course are the subjects of *maps* of distributive laws. We speculate the existence of *Kleisli* maps of distributive laws that guarantee preservation of equivalences weaker than bisimilarity. We plan to develop this notion and explore its applicability in future work.

Conclusion. It is evident that the systematic approach presented in this work may markedly streamline proofs for compiler security as it involves a single, simple coherence criterion. Explicit reasoning about program contexts is no longer necessary, but that does not mean that contexts are irrelevant. On the contrary, the guarantees are implicitly *contextual* due to the well-behavedness of the semantics. Finally, while the overall usability and eventual success of our method remains a question mark as it depends on the expressiveness of the threat model, the body of work in coalgebraic weak bisimilarity and distributive laws in Kleisli categories suggests that there are many promising avenues for further progress.

Acknowledgements. This work was partially supported by the Research Fund KU Leuven. Andreas Nuyts holds a PhD fellowship from the Research Foundation - Flanders (FWO).

References

1. Abadi, M.: Protection in programming-language translations. In: Secure Internet Programming, Security Issues for Mobile and Distributed Objects, pp. 19–34 (1999). https://doi.org/10.1007/3-540-48749-2_2
2. Abate, C., et al.: Journey beyond full abstraction: exploring robust property preservation for secure compilation (2018). arXiv: 1807.04603 [cs.PL]
3. Abate, C., et al.: When good components go bad: formally secure compilation despite dynamic compromise. In: Lie, D. et al. (ed.) Proceedings of the 2018 ACM SIGSAC Conference on Computer and Communications Security, CCS 2018, Toronto, ON, Canada, 15–19 October 2018, pp. 1351–1368 (2018). https://doi.org/10.1145/3243734.3243745, ISBN: 978-1-4503-5693-0
4. Abbott, M.G., Altenkirch, T., Ghani, N.: Containers: constructing strictly positive types. Theor. Comput. Sci. **342**(1), 3–27 (2005). https://doi.org/10.1016/j.tcs.2005.06.002
5. Abbott, M.G., et al.: For data: differentiating data structures. Fundam. Inform. 65(1-2), 1–28 (2005). http://content.iospress.com/articles/fundamenta-informaticae/fi65-1-2-02
6. Abou-Saleh, F.: A coalgebraic semantics for imperative programming languages. PhD thesis. Imperial College London, UK (2014). http://hdl.handle.net/10044/1/13693
7. Abou-Saleh, F., Pattinson, D.: Towards effects in mathematical operational semantics. Electr. Notes Theor. Comput. Sci. **276**, 81–104 (2011). https://doi.org/10.1016/j.entcs.2011.09.016

8. Agten, P., et al.: Secure compilation to modern processors. In: Chong, S. (ed.) 25th IEEE Computer Security Foundations Symposium, CSF 2012, Cambridge, MA, USA, 25–27 June 2012. IEEE Computer Society, pp. 171–185 (2012). https://doi. org/10.1109/CSF.2012.12, ISBN: 978-1-4673-1918-8

9. Ahmed, A., Blume, M.: An equivalence-preserving CPS translation via multi-language semantics. In: Chakravarty, M.M.T., Hu, Z., Danvy, O. (eds.) Proceeding of the 16th ACM SIGPLAN International Conference on Functional Programming, ICFP 2011, Tokyo, Japan, 19–21 September 2011, pp. 431–444. ACM (2011). https://doi.org/10.1145/2034773.2034830, ISBN: 978-1-4503-0865-6

10. Ahmed, A., et al.: Secure compilation (Dagstuhl Seminar 18201). In: Ahmed, A., et al. (ed.) Dagstuhl Reports 8.5 (2018), pp. 1–30. ISSN: 2192–5283.https://doi. org/10.4230/DagRep.8.5.1, http://drops.dagstuhl.de/opus/volltexte/2018/9891

11. Bartels, F.: On generalised coinduction and probabilistic specification formats: distributive laws in coalgebraic modelling (2004)

12. Bonchi, F., et al.: Lax bialgebras and up-to techniques for weak bisimulations. In: Aceto, L., de Frutos-Escrig, D. (eds.) 26th International Conference on Concurrency Theory, CONCUR 2015, Madrid, Spain, September 1.4, 2015, vol. 42. LIPIcs. Schloss Dagstuhl - Leibniz-Zentrum fuer Informatik, pp. 240–253 (2015). https://doi.org/10.4230/LIPIcs.CONCUR.2015.240, ISBN: 978-3-939897-91-0

13. Brengos, T.: Weak bisimulation for coalgebras over order enriched monads. Logical Methods Comput. Sci. $11(2)$ (2015). https://doi.org/10.2168/LMCS-11(2:14)2015

14. Cockett, J.R.B.: Introduction to distributive categories. In: Math. Struct. Comput. Sci. 33, 277–307 (1993) . https://doi.org/10.1017/S0960129500000232

15. Costan, V., Devadas, S.: Intel SGX explained. In: IACR Cryptology ePrint Archive 2016 (2016), p. 86. http://eprint.iacr.org/2016/086

16. Devriese, D., Patrignani, M., Piessens, F.: Fully-abstract compilation by approximate back-translation. In: Proceedings of the 43rd Annual ACM SIGPLAN-SIGACT Symposium on Principles of Programming Languages, POPL 2016, St. Petersburg, FL, USA, 20–22 January 2016, pp. 164–177 (2016). https://doi.org/ 10.1145/2837614.2837618

17. Dreyer, D., Ahmed, A., Birkedal, L.: Logical step-indexed logical relations. Logical Methods Comput. Sci. $7(2)$ (2011). https://doi.org/10.2168/LMCS-7(2:16)2011

18. Fournet, C., et al.: Fully abstract compilation to JavaScript. In: The 40th Annual ACM SIGPLAN-SIGACT Symposium on Principles of Programming Languages, POPL 2013, Rome, Italy - January 23–25, 2013, pp. 371–384 (2013). https://doi. org/10.1145/2429069.2429114

19. Groote, J.F., Vaandrager, F.W.: Structured operational semantics and bisimulation as a congruence. Inf. Comput. $100(2)$, 202–260 (1992). https://doi.org/10.1016/ 0890-5401(92)90013-6

20. Hasuo, I., Jacobs, B., Sokolova, A.: Generic trace semantics via coinduction. Logical Methods Comput. Sci. $3(4)$ (2007). https://doi.org/10.2168/LMCS-3(4:11)2007

21. Intel 64 and IA-32 Architectures Software Developer's Manual. Intel Corporation (2016). https://www.intel.com/content/dam/www/public/us/en/documents/ manuals/64-ia-32-architectures-software-developerinstruction-set-reference-manual-325383.pdf

22. Jacobs, B.: Introduction to coalgebra: towards mathematics of states and observation, vol. 59. Cambridge Tracts in Theoretical Computer Science. Cambridge University Press (2016). ISBN: 9781316823187. CBO9781316823187. https://doi. org/10.1017/CBO9781316823187

23. Jacobs, B.: Parameters and parametrization in specification, using distributive categories. In: Fundam. Inform. **24**(3), 209–250 (1995). https://doi.org/10.3233/FI-1995-2431

24. Jagadeesan, R., et al.: Local memory via layout randomization. In: Proceedings of the 24th IEEE Computer Security Foundations Symposium, CSF 2011, Cernay-la-Ville, France, 27–29 June 2011, pp. 161–174. IEEE Computer Society, (2011). ISBN: 978-1-61284-644-6. https://doi.org/10.1109/CSF.2011.18

25. Kennedy, A.: Securing the .NET programming model. In: Theor. Comput. Sci. **364**(3), 311–317 (2006). https://doi.org/10.1016/j.tcs.2006.08.014

26. Klin, B.: Bialgebras for structural operational semantics: an introduction. Theor. Comput. Sci. **412**(38), 5043–5069 (2011). https://doi.org/10.1016/j.tcs.2011.03.023

27. Klin, B., Nachyla, B.: Presenting morphisms of distributive laws. In: 6th Conference on Algebra and Coalgebra in Computer Science, CALCO 2015, 24–26 June 2015, Nijmegen, The Netherlands, pp. 190–204 (2015). https://doi.org/10.4230/LIPIcs.CALCO.2015.190

28. Lenisa, M., Power, J., Watanabe, H.: Distributivity for endofunctors, pointed and co-pointed endofunctors, monads and comonads. Electr. Notes Theor. Comput. Sci. **33**, 230–260 (2000). https://doi.org/10.1016/S1571-0661(05)80350-0

29. Mcbride, C.: The derivative of a regular type is its type of one-hole contexts (Extended Abstract) (2001)

30. MIPS Architecture for Programmers Volume II-A: The MIPS32 Instruction Set Manual. MIPS Technologies (2016). https://s3-eu-west-1.amazonaws.com/downloads-mips/documents/MD00086-2B-MIPS32BISAFP-6.06.pdf

31. Morris, J.H.: Lambda-calculus models of programming languages. PhD thesis. Massachusetts Institute of Technology (1968)

32. New, M.S., Bowman, W.J., Ahmed, A.: Fully abstract compilation via universal embedding. In: Garrigue, J., Keller, G., Sumii, E. (eds.) Proceedings of the 21st ACM SIGPLAN International Conference on Functional Programming, ICFP 2016, Nara, Japan, 18–22 September 2016, pp. 103–116. ACM (2016). ISBN: 978-1-4503-4219- 3. https://doi.org/10.1145/2951913.2951941

33. Patrignani, M., Ahmed, A., Clarke, D.: Formal approaches to secure compilation: a survey of fully abstract compilation and related work. ACM Comput. Surv. **51**(6), 125:1–125:36 (2019). https://doi.org/10.1145/3280984, ISSN: 0360-0300

34. Patrignani, M., Clarke, D., Piessens, F.: Secure compilation of object-oriented components to protected module architectures. In: Shan, C. Programming Languages and Systems - 11th Asian Symposium, APLAS 2013, Melbourne, VIC, Australia, 9–11 December 2013. Proceedings, vol. 8301. Lecture Notes in Computer Science, pp. 176–191. Springer, Heidelberg (2013). https://doi.org/10.1007/978-3-319-03542-0_13, ISBN: 978-3-319-03541-3

35. Patrignani, M., Devriese, D., Piessens, F.: On modular and fully-abstract compilation. In: IEEE 29th Computer Security Foundations Symposium, CSF 2016, Lisbon, Portugal, June 27 - July 1, 2016, pp. 17–30. IEEE Computer Society (2016). ISBN: 978-1-5090-2607-4. https://doi.org/10.1109/CSF.2016.9

36. Patrignani, M., Garg, D.: Robustly safe compilation. In: Programming Languages and Systems - 28th European Symposium on Programming, ESOP 2019, Held as Part of the European Joint Conferences on Theory and Practice of Software, ETAPS 2019, Prague, Czech Republic, 6–11 April 2019, Proceedings, pp. 469–498 (2019). https://doi.org/10.1007/978-3-030-17184-1_17

37. Patrignani, M., et al.: Secure compilation to protected module architectures. ACM Trans. Program. Lang. Syst. **37**(2), 6:1–6:50 (2015). https://doi.org/10.1145/2699503
38. Plotkin, G.D.: A structural approach to operational semantics. J. Log. Algebr. Program. **60–61**, 17–139 (2004)
39. Popescu, A.: Weak bisimilarity coalgebraically. In: Algebra and Coalgebra in Computer Science, Third International Conference, CALCO 2009, Udine, Italy, 7–10 September 2009. Proceedings, pp. 157–172 (2009). https://doi.org/10.1007/978-3-642-03741-2_12
40. Power, J., Watanabe, H.: Distributivity for a monad and a comonad. Electr. Notes Theor. Comput. Sci. **19**, 102 (1999). https://doi.org/10.1016/S1571-0661(05)80271-3
41. Rot, J., et al.: Enhanced coalgebraic bisimulation. Math. Struct. Comput. Sci. **27**(7), 1236–1264 (2017). https://doi.org/10.1017/S0960129515000523
42. Rothe, J., Masulovic, D.: Towards weak bisimulation for coalgebras. Electr. Notes Theor. Comput. Sci. **68**(1), 32–46 (2002). https://doi.org/10.1016/S1571-0661(04)80499-7
43. Rutten, J.J.M.M.: A note on coinduction and weak bisimilarity for while programs. In: ITA 33.4/5, pp. 393–400 (1999). https://doi.org/10.1051/ita:1999125
44. Skorstengaard, L., Devriese, D., Birkedal, L.: Reasoning about a machine with local capabilities. In: Ahmed, A. (ed.) ESOP 2018. LNCS, vol. 10801, pp. 475–501. Springer, Cham (2018). https://doi.org/10.1007/978-3-319-89884-1_17
45. Skorstengaard, L., Devriese, D., Birkedal, L.: StkTokens: enforcing well-bracketed control flow and stack encapsulation using linear capabilities. Proc. ACM Program. Lang. **3**(POPL), 19:1–19:28 (2019). https://doi.org/10.1145/3290332, ISSN: 2475-1421
46. Tsampas, S., Devriese, D., Piessens, F.: Temporal safety for stack allocated memory on capability machines. In: 32nd IEEE Computer Security Foundations Symposium, CSF 2019, Hoboken, NJ, USA, 25–28 June 2019, pp. 243–255. IEEE (2019). https://doi.org/10.1109/CSF.2019.00024, ISBN: 978-1-7281-1407-1
47. Turi, D.: Categorical modelling of structural operational rules: case studies. In: Category Theory and Computer Science, 7th International Conference, CTCS '97, Santa Margherita Ligure, Italy, 4–6 September 1997, Proceedings, pp. 127–146 (1997). https://doi.org/10.1007/BFb0026985
48. Turi, D., Plotkin, G.D.: Towards a mathematical operational semantics. In: Proceedings, 12th Annual IEEE Symposium on Logic in Computer Science, Warsaw, Poland, June 29 - July 2, 1997, pp. 280–291 (1997). https://doi.org/10.1109/LICS.1997.614955
49. Van Strydonck, T., Piessens, F., Devriese, D.: Linear capabilities for fully abstract compilation of separation-logic-verified code. Proc. ACM Program. Lang. ICFP (2019). accepted
50. Watanabe, H.: Well-behaved translations between structural operational semantics. Electr. Notes Theor. Comput. Sci. **65**(1), 337–357 (2002). https://doi.org/10.1016/S1571-0661(04)80372-4
51. Watson, R.N.M., et al.: CHERI: a hybrid capability-system architecture for scalable software compartmentalization. In: 2015 IEEE Symposium on Security and Privacy, SP 2015, San Jose, CA, USA, 17–21 May 2015, pp. 20–37. IEEE Computer Society (2015). https://doi.org/10.1109/SP.2015.9, ISBN: 978-1-4673-6949-7

Semantics for First-Order Affine Inductive Data Types via Slice Categories

Vladimir Zamdzhiev[(✉)]

Université de Lorraine, CNRS, Inria, LORIA, 54000 Nancy, France
vladimir.zamdzhiev@loria.fr

Abstract. Affine type systems are substructural type systems where copying of information is restricted, but discarding of information is permissible at all types. Such type systems are well-suited for describing quantum programming languages, because copying of quantum information violates the laws of quantum mechanics. In this paper, we consider a first-order affine type system with inductive data types and present a novel categorical semantics for it. The most challenging aspect of this interpretation comes from the requirement to construct appropriate discarding maps for our data types which might be defined by mutual/nested recursion. We show how to achieve this for all types by taking models of a first-order linear type system whose atomic types are discardable and then presenting an additional affine interpretation of types within the slice category of the model with the tensor unit. We present some concrete categorical models for the language ranging from classical to quantum. Finally, we discuss potential ways of dualising and extending our methods and using them for interpreting coalgebraic and lazy data types.

Keywords: Inductive data types · Categorical semantics · Affine types

1 Introduction

Linear Logic [5] is a substructural logic where the rules for weakening and contraction are restricted. Linear logic has been very influential in computer science and has lead to the development of linear type systems where discarding and copying of variables is restricted. Linear logic has also inspired the development of *affine* type systems, which are substructural type systems where only the rule for contraction (copying of variables) is restricted, but weakening (discarding of variables) is completely unrestricted. Affine type systems are a natural choice for quantum programming languages [2,17,19,20], because they can be used to enforce compliance with the laws of quantum mechanics, where copying of quantum information is impossible [23].

In this paper we consider a first-order affine type system with inductive data types, called **Aff**, and we present a categorical semantics for it. The main focus

© IFIP International Federation for Information Processing 2020
Published by Springer Nature Switzerland AG 2020
D. Petrişan and J. Rot (Eds.): CMCS 2020, LNCS 12094, pp. 180–200, 2020.
https://doi.org/10.1007/978-3-030-57201-3_10

of the present paper is on the construction of the required discarding maps that are necessary for the interpretation of the type system. Our semantics is novel in that we assume very little structure on the model side: we do not assume the existence of any (sub)category where the tensor unit I is a terminal object (e.g. [6]). Instead, we merely assume that the interpretation of every *atomic* type is equipped with some discarding map (which is clearly necessary) and we then show how to construct *all other* discarding maps by providing an *affine interpretation of types* within the slice category of the model with the tensor unit. Thus, by taking a categorical model of a first-order linear type system, we construct all the discarding maps we need by performing a careful *semantic* analysis, instead of assuming additional structure within the categorical model.

Outline. We begin by recalling some background about parameterised initial algebras in Sect. 2. Next, we describe the syntax of **Aff**, which is a fragment of the quantum programming language QPL [16,17,19], in Sect. 3. In Sect. 4, we present the operational semantics of **Aff**. One of our contributions is in Sect. 5, where we show how parameterised initial algebras for suitable functors may be reflected into slice categories. Our main contributions are in Sect. 6, where we describe a categorical model for our language, and in Sect. 7, where we present a novel categorical semantics for the affine structure of types by providing a non-standard type interpretation within a slice category. In Sect. 8 we discuss future work and possible extensions and in Sect. 9 we discuss related work and present some concluding remarks.

2 Parameterised Initial Algebras

Simple inductive data types, like lists and natural numbers, may be interpreted by initial algebras. However, the interpretation of inductive data types defined by mutual/nested induction requires a more general notion called *parameterised initial algebra*, which we shall now recall.

Definition 1. (cf. [4, §6.1]). *Let* **A** *and* **B** *be categories and* $T : \mathbf{A} \times \mathbf{B} \to \mathbf{B}$ *a functor. A* parameterised initial algebra *for* T *is a pair* $(T^\dagger, \tau)$, *such that:*

- $T^\dagger : \mathbf{A} \to \mathbf{B}$ *is a functor;*
- $\tau : T \circ \langle Id, T^\dagger \rangle \Rightarrow T^\dagger : \mathbf{A} \to \mathbf{B}$ *is a natural isomorphism;*
- *For every* $A \in \mathrm{Ob}(\mathbf{A})$, *the pair* $(T^\dagger A, \tau_A)$ *is an initial* $T(A, -)$-*algebra.*

Note that by trivialising **A**, we get the well-known notion of initial algebra. Next, we recall a theorem which provides sufficient conditions for the existence of parameterised initial algebras.

Theorem 2. ([13, Theorem 4.12]). *Let* **B** *be a category with an initial object and all* ω-*colimits. Let* $T : \mathbf{A} \times \mathbf{B} \to \mathbf{B}$ *be an* ω-*cocontinuous functor. Then* T *has a parameterised initial algebra* $(T^\dagger, \tau)$ *and the functor* $T^\dagger$ *is also* ω-*cocontinuous.*

In particular, the above theorem shows that ω-cocontinuous functors are closed under formation of parameterised initial algebras.

Type Variables	X, Y, Z	
Term Variables	x, y, b, u	
Atomic Types	$\mathbf{A} \in \mathcal{A}$	
Types	$A, B, C ::=$	$X \mid I \mid \mathbf{A} \mid A + B \mid A \otimes B \mid \mu X.A$
Terms	$M, N ::=$	**new unit** u | **discard** x | $M; N$ | **skip** |

$$\textbf{while } b \textbf{ do } M \mid x = \textbf{left}_{A,B} M \mid x = \textbf{right}_{A,B} M \mid$$
$$\textbf{case } y \textbf{ of } \{\textbf{left } x_1 \to M \mid \textbf{right } x_2 \to N\} \mid$$
$$x = (x_1, x_2) \mid (x_1, x_2) = x \mid y = \textbf{fold } x \mid y = \textbf{unfold } x$$

Type contexts	Θ	$::=$	$X_1, X_2, \ldots, X_n$
Variable contexts Γ, Σ		$::=$	$x_1 : A_1, \ldots, x_n : A_n$
Type Judgements $\Theta \vdash A$			
Term Judgements $\vdash \langle \Gamma \rangle M \langle \Sigma \rangle$			

Fig. 1. Syntax of **Aff**.

3 Syntax of Aff

In this section we describe the syntax of **Aff**, which is the language on which we will base the development of our ideas. **Aff** is a fragment of the quantum programming language QPL [17] which is obtained from QPL by removing procedures, quantum resources and copying of classical information. The reason for considering this fragment is just simplicity and brevity of the presentation.

Remark 3. In fact, the methods we describe can handle the addition of procedures and the copying of non-linear information with no further effort. The addition of quantum resources can also be handled by our methods, but this requires identifying a suitable category of quantum computation with ω-colimits.

The syntax of **Aff** is summarised in Fig. 1. A type context Θ, is *well-formed*, denoted $\vdash \Theta$, if Θ is simply a list of distinct type variables. Well-formed types, denoted $\Theta \vdash A$, are specified by the following rules:

$$\frac{\vdash \Theta}{\Theta \vdash \Theta_i} \quad \frac{\vdash \Theta}{\Theta \vdash I} \quad \frac{\vdash \Theta}{\Theta \vdash \mathbf{A}} \quad \frac{\Theta \vdash A \quad \Theta \vdash B}{\Theta \vdash A \star B} \star \in \{+, \otimes\} \quad \frac{\Theta, X \vdash A}{\Theta \vdash \mu X.A} ,$$

where **A** ranges over a set of atomic types $\mathcal{A}$, which we will leave unspecified in this paper (for generality). For example, in quantum programming, it suffices to assume $\mathcal{A} = \{\mathbf{qubit}\}$. This is the case for QPL.

A type A is *closed* whenever $\cdot \vdash A$. Note that nested type induction (also known as mutual induction) is allowed, i.e., it is possible to form inductive data types which have more than one free variable in their type contexts. Henceforth, we implicitly assume that all types we are dealing with are well-formed.

Example 4. Natural numbers can be defined as $\mathbf{Nat} \equiv \mu X.I + X$. A list of a closed type $\cdot \vdash A$ is defined by $\mathbf{List}(A) \equiv \mu Y.I + A \otimes Y$.

Term variables are denoted by small Latin characters (e.g. x, y, u, b). In particular, u ranges over variables of unit type I, b over variables of type $\mathbf{bit} := I + I$

$$\overline{\vdash \langle \Gamma \rangle \textbf{ new unit } u \ \langle \Gamma, u : I \rangle} \qquad \overline{\vdash \langle \Gamma, x : A \rangle \textbf{ discard } x \ \langle \Gamma \rangle}$$

$$\overline{\vdash \langle \Gamma \rangle \textbf{ skip } \langle \Gamma \rangle}$$

$$\frac{\vdash \langle \Gamma \rangle \ M \ \langle \Gamma' \rangle \qquad \vdash \langle \Gamma' \rangle \ N \ \langle \Sigma \rangle}{\vdash \langle \Gamma \rangle \ M; N \ \langle \Sigma \rangle} \qquad \frac{\vdash \langle \Gamma, b : \textbf{bit} \rangle \ M \ \langle \Gamma, b : \textbf{bit} \rangle}{\vdash \langle \Gamma, b : \textbf{bit} \rangle \textbf{ while } b \textbf{ do } M \ \langle \Gamma, b : \textbf{bit} \rangle}$$

$$\overline{\vdash \langle \Gamma, x : A \rangle \ y = \textbf{left}_{A,B} \ x \ \langle \Gamma, y : A + B \rangle}$$

$$\overline{\vdash \langle \Gamma, x : B \rangle \ y = \textbf{right}_{A,B} \ x \ \langle \Gamma, y : A + B \rangle}$$

$$\frac{\vdash \langle \Gamma, x_1 : A \rangle \ M_1 \ \langle \Sigma \rangle \qquad \vdash \langle \Gamma, x_2 : B \rangle \ M_2 \ \langle \Sigma \rangle}{\vdash \langle \Gamma, y : A + B \rangle \textbf{ case } y \textbf{ of } \{\textbf{left}_{A,B} \ x_1 \rightarrow M_1 \mid \textbf{right}_{A,B} \ x_2 \rightarrow M_2 \ \} \ \langle \Sigma \rangle}$$

$$\overline{\vdash \langle \Gamma, x_1 : A, x_2 : B \rangle \ x = (x_1, x_2) \ \langle \Gamma, x : A \otimes B \rangle}$$

$$\overline{\vdash \langle \Gamma, x : A \otimes B \rangle \ (x_1, x_2) = x \ \langle \Gamma, x_1 : A, x_2 : B \rangle}$$

$$\overline{\vdash \langle \Gamma, x : A[\mu X.A/X] \rangle \ y = \textbf{fold}_{\mu X.A} \ x \ \langle \Gamma, y : \mu X.A \rangle}$$

$$\overline{\vdash \langle \Gamma, x : \mu X.A \rangle \ y = \textbf{unfold } x \ \langle \Gamma, y : A[\mu X.A/X] \rangle}$$

Fig. 2. Formation rules for **Aff** terms.

and x, y range over arbitrary variables. *Variable contexts* are denoted by capital Greek letters, such as Γ and Σ. Variable contexts contain only variables of closed types and are written as $\Gamma = x_1 : A_1, \ldots, x_n : A_n$.

A *term judgement* $\vdash \langle \Gamma \rangle M \langle \Sigma \rangle$ indicates that term M is well-formed assuming an input variable context Γ and an output variable context Σ. Note that, the input/output contexts here simply describe the variables before/after execution of a program and they do not refer to any sort of I/O interactivity. All types within a term judgement are necessarily closed. The formation rules are shown in Fig. 2.

The term **new unit** u declares a new variable u of unit type which may then be used to define more complicated terms. The term **discard** x is of central importance in this paper, because it allows us to discard any variable of any type which has been defined. For example, the following program:

$$\vdash \langle \cdot \rangle \textbf{ new unit } u; \ l = \textbf{left}_{I,\textbf{Nat}} \ u; \ zero = \textbf{fold}_{\textbf{Nat}} \ l; \textbf{ discard } zero \ \langle \cdot \rangle$$

defines a variable *zero* which represents the zero natural number and then immediately discards it (without using the elimination rules for inductive types and sum types). In fact, this program is equivalent to $\vdash \langle \cdot \rangle \textbf{ skip } \langle \cdot \rangle$.

Remark 5. Because we are not concerned with any domain-specific applications in this paper, we leave the atomic types uninhabited. Of course, any domain-specific extension should add suitable introduction and elimination rules for each atomic type. In the case of QPL, the term language has to be extended with three terms – one each for preparing a qubit in state $|0\rangle$, applying a unitary gate to a term and finally measuring a qubit. See [17] for more information.

4   Operational Semantics of Aff

The purpose of this section is to present the operational semantics of **Aff**. We begin by introducing *program configurations* which completely and formally describe the current state of program execution. A program configuration is a pair $(M \mid V)$, where M is the term which remains to be executed and V is a *value assignment*, which is a function that assigns values to variables that have already been introduced.

Value Assignments. *Values* are expressions defined by the following grammar:

$$v, w ::= * \mid \mathbf{left}_{A,B} v \mid \mathbf{right}_{A,B} v \mid (v, w) \mid \mathbf{fold}_{\mu X.A} v.$$

The expression $*$ represents the unique value of unit type I. Other particular values of interest are the canonical values of type **bit**, called *false* and *true*, which are formally defined by $\mathbf{ff} := \mathbf{left}_{I,I} *$ and $\mathbf{tt} := \mathbf{right}_{I,I} *$. They play an important role in the operational semantics.

The well-formed values, denoted $\vdash v : A$, are specified by the following rules:

$$\frac{}{\vdash * : I} \qquad \frac{\vdash v : A}{\vdash \mathbf{left}_{A,B} v : A + B} \qquad \frac{\vdash v : B}{\vdash \mathbf{right}_{A,B} v : A + B}$$

$$\frac{\vdash v : A \qquad \vdash w : B}{\vdash (v, w) : A \otimes B} \qquad \frac{\vdash v : A[\mu X.A/X]}{\vdash \mathbf{fold}_{\mu X.A} v : \mu X.A}$$

A *value assignment* is simply a function from term variables to values. We write value assignments as $V = \{x_1 = v_1, \ldots, x_n = v_n\}$, where each x_i is a variable and each v_i is a value. We say that V is *well-formed* in variable context $\Gamma = \{x_1 : A_1, \ldots x_n : A_n\}$, denoted $\Gamma \vdash V$, if V has the same variables as Γ and $\vdash v_i : A_i$, for each $i \in \{1, \ldots n\}$.

Program Configurations. A *program configuration* is a couple $(M \mid V)$, where M is a term and where V is a value assignment. A *well-formed* program configuration, denoted $\Gamma; \Sigma \vdash (M \mid V)$, is a program configuration $(M \mid V)$, such that there exist (necessarily unique) Γ, Σ with: (1) $\vdash \langle \Gamma \rangle M \langle \Sigma \rangle$ is a well-formed term; and (2) $\Gamma \vdash V$ is a well-formed value assignment.

The (small step) operational semantics is defined as a relation[1] $(- \rightsquigarrow -)$ on program configurations $(M \mid V)$ by induction on the structure of M in Fig. 3. Note that, in the rule for while loops, the term **if** b **then** $\{M\}$ is just syntactic sugar for **case** b **of** $\{\mathbf{left}\ u \rightarrow b = \mathbf{left}\ u \mid \mathbf{right}\ u \rightarrow b = \mathbf{right}\ u; M\ \}$.

---

[1] In fact, this relation is a partial function.

$$(\textbf{new unit } u \mid V) \rightsquigarrow (\textbf{skip} \mid V, u = *)$$

$$(\textbf{discard } x \mid V, x = v) \rightsquigarrow (\textbf{skip} \mid V)$$

$$(\textbf{skip}; P \mid V) \rightsquigarrow (P \mid V)$$

$$\frac{(P \mid V) \rightsquigarrow (P' \mid V')}{(P; Q \mid V) \rightsquigarrow (P'; Q \mid V')}$$

$$(\textbf{while } b \textbf{ do } M \mid V, b = v) \rightsquigarrow (\textbf{if } b \textbf{ then } \{M; \textbf{while } b \textbf{ do } M\} \mid V, b = v)$$

$$(y = \textbf{left } x \mid V, x = v) \rightsquigarrow (\textbf{skip} \mid V, y = \textbf{left } v)$$

$$(y = \textbf{right } x \mid V, x = v) \rightsquigarrow (\textbf{skip} \mid V, y = \textbf{right } v)$$

$$(\textbf{case } y \textbf{ of } \{\textbf{left } x_1 \rightarrow M_1 \mid \textbf{right } x_2 \rightarrow M_2 \} \mid V, y = \textbf{left } v) \rightsquigarrow (M_1 \mid V, x_1 = v)$$

$$(\textbf{case } y \textbf{ of } \{\textbf{left } x_1 \rightarrow M_1 \mid \textbf{right } x_2 \rightarrow M_2 \} \mid V, y = \textbf{right } v) \rightsquigarrow (M_2 \mid V, x_2 = v)$$

$$(x = (x_1, x_2) \mid V, x_1 = v_1, x_2 = v_2) \rightsquigarrow (\textbf{skip} \mid V, x = (v_1, v_2))$$

$$((x_1, x_2) = x \mid V, x = (v_1, v_2)) \rightsquigarrow (\textbf{skip} \mid V, x_1 = v_1, x_2 = v_2)$$

$$(y = \textbf{fold } x \mid V, x = v) \rightsquigarrow (\textbf{skip} \mid V, y = \textbf{fold } v)$$

$$(y = \textbf{unfold } x \mid V, x = \textbf{fold } v) \rightsquigarrow (\textbf{skip} \mid V, y = v)$$

Fig. 3. Small step operational semantics of **Aff**.

Theorem 6. (Subject reduction [17]). *If $\Gamma; \Sigma \vdash (M \mid V)$ and $(M \mid V) \rightsquigarrow (M', V')$, then $\Gamma'; \Sigma \vdash (M', V')$, for some (necessarily unique) context Γ'.*

Assumption 7. *Henceforth, all configurations are assumed to be well-formed.*

We shall use calligraphic letters $(\mathcal{C}, \mathcal{D}, \ldots)$ to denote configurations. A *terminal* configuration is a configuration $\mathcal{C}$, such that $\mathcal{C} = (\textbf{skip}, V)$.

Theorem 8. (Progress [17]). *If $\mathcal{C}$ is a configuration, then either $\mathcal{C}$ is terminal or there exists a configuration $\mathcal{D}$, such that $\mathcal{C} \rightsquigarrow \mathcal{D}$.*

Remark 9. Any domain-specific extension should, of course, also adapt the operational semantics as necessary. In the case of QPL, this requires introducing new reduction rules for the additional terms and extending the notion of configuration with an extra component that stores the quantum data.

5   Slice Categories for Affine Types

Our type system is affine and in order to provide a denotational interpretation we have to construct discarding maps in our model at every type. This is achieved in the following way: (1) for every closed type A we provide a standard interpretation $[\![A]\!] \in \mathrm{Ob}(\mathbf{C})$ in our model $\mathbf{C}$; (2) in addition, we provide a *affine* type interpretation $[\![A]\!] \in \mathrm{Ob}(\mathbf{C}/I)$ within the *slice category with the tensor unit*, that is, for every type we carefully pick out a specific discarding map; (3) we prove $[\![A]\!] = ([\![A]\!], \diamond_A : [\![A]\!] \to I)$ and show that our choice of discarding map $\diamond_A$ can discard all values of our language, as required.

The purpose of this section is to show the slice category $\mathbf{C}/I$ has sufficient categorical structure for the affine interpretation of types. Our analysis is quite general and works for many affine scenarios. Under some basic assumptions on $\mathbf{C}$ we show that $\mathbf{C}/I$ inherits from $\mathbf{C}$: a symmetric monoidal structure (Proposition 13), finite coproducts (Proposition 14) and (parameterised) initial algebras for a sufficiently large class of functors (Theorem 18).

Assumption 10. *Throughout the remainder of the section we assume we are given a category $\mathbf{C}$ and we fix an object $I \in \mathrm{Ob}(\mathbf{C})$. Let $\mathbf{C}_a := \mathbf{C}/I$ be the slice category of $\mathbf{C}$ with the fixed object I.*

Thus, the objects of $\mathbf{C}_a$ are pairs $(A, \diamond_A)$, where $A \in \mathrm{Ob}(\mathbf{C})$ and $\diamond_A : A \to I$ is a morphism of $\mathbf{C}$. Then, a morphism $f : (A, \diamond_A) \to (B, \diamond_B)$ of $\mathbf{C}_a$ is a morphism $f : A \to B$ of $\mathbf{C}$, such that $\diamond_B \circ f = \diamond_A$. Composition and identities are the same as in $\mathbf{C}$. We refer to the maps $\diamond_A$ as the *discarding* maps and to the morphisms of $\mathbf{C}_a$ as *affine* maps.

Notation 11. *There exists an obvious forgetful functor $U : \mathbf{C}_a \to \mathbf{C}$ given by $U(A, \diamond_A) = A$ and $U(f) = f$.*

The following (well-known) proposition will be used to show the existence of certain initial algebras in $\mathbf{C}_a$.

Proposition 12. *The functor $U : \mathbf{C}_a \to \mathbf{C}$ reflects small colimits.*

Proof. Straightforward verification.   □

Our next (well-known) proposition shows how a symmetric monoidal structure on $\mathbf{C}$ induces one on $\mathbf{C}_a$.

Proposition 13. *Assume that $\mathbf{C}$ is equipped with a (symmetric) monoidal structure $(\mathbf{C}, \otimes, I, \alpha, \lambda, \rho, (\sigma))$. Then, the tuple $(\mathbf{C}_a, \otimes_a, (I, id_I), \alpha_a, \lambda_a, \rho_a, (\sigma_a))$ is a (symmetric) monoidal category, where $\otimes_a : \mathbf{C}_a \times \mathbf{C}_a \to \mathbf{C}_a$ is defined by:*

$$(A, \diamond_A) \otimes_a (B, \diamond_B) := (A \otimes B, \lambda_I \circ (\diamond_A \otimes \diamond_B))$$

$$f \otimes_a g := f \otimes g$$

and where the natural isomorphisms $\alpha_a, \lambda_a, \rho_a, (\sigma_a)$ are componentwise equal to $\alpha, \lambda, \rho, (\sigma)$ respectively. Moreover, this data makes $U : \mathbf{C}_a \to \mathbf{C}$ a strict monoidal functor and we also have:

$$\otimes \circ (U \times U) = U \circ \otimes_a : \mathbf{C}_a \times \mathbf{C}_a \to \mathbf{C}.$$

Proof. Straightforward verification.   □

The next (well-known) proposition shows how coproducts on $\mathbf{C}$ induce coproducts on $\mathbf{C}_a$.

Proposition 14. *Assume that $\mathbf{C}$ has finite coproducts with initial object denoted $\varnothing$ and binary coproducts by $(A+B, left_{A,B}, right_{A,B})$. Then, the category $\mathbf{C}_a$ has finite coproducts. Its initial object is $(\varnothing, \perp_{\varnothing,I})$ and binary coproducts are given by $(A, \diamond_A) +_a (B, \diamond_B) := (A + B, [\diamond_A, 0_B])$. Moreover, we have:*

$$+ \circ (U \times U) = U \circ +_a : \mathbf{C}_a \times \mathbf{C}_a \to \mathbf{C}.$$

Proof. Straightforward verification.   □

5.1 (Parameterised) Initial Algebras in $\mathbf{C}_a$

In this subsection we will show how (parameterised) initial algebras from $\mathbf{C}$ may be reflected into $\mathbf{C}_a$ by using methods from [10,13]. Towards this end, we assume that $\mathbf{C}$ has some additional structure, so that parameterised initial algebras may be formed within it.

Assumption 15. *Throughout the remainder of the section, we assume that $\mathbf{C}$ has an initial object $\varnothing$ and all ω-colimits.*

We begin by recalling how colimits in $\mathbf{C}$ induce colimits in $\mathbf{C}_a$.

Proposition 16. *The category $\mathbf{C}_a$ has an initial object and all ω-colimits. Moreover, the forgetful functor $U : \mathbf{C}_a \to \mathbf{C}$ preserves and reflects ω-colimits.*

Proof. This fact is well-known and it follows immediately by examining the proof of [1, Proposition 2.16.3(2)].   □

Next, we show that the functor U may be used to reflect ω-cocontinuity of functors on $\mathbf{C}$ to functors on $\mathbf{C}_a$.

Theorem 17. *Let $H : \mathbf{C}_a^n \to \mathbf{C}_a$ be a functor and $T : \mathbf{C}^n \to \mathbf{C}$ an ω-cocontinuous functor, such that the diagram:*

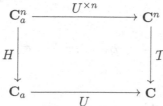

commutes. Then, H is also ω-cocontinuous.

Proof. Let $D : \omega \to \mathbf{C}_a^n$ be an arbitrary ω-diagram in $\mathbf{C}_a^n$ and let μ be its colimiting cocone. Since U preserves ω-colimits (Proposition 16), it follows that $U^{\times n}\mu$ is a colimiting cocone of $U^{\times n}D$ in $\mathbf{C}^n$. By assumption T is ω-cocontinuous, so $TU^{\times n}\mu$ is a colimiting cocone of $TU^{\times n}D$ in $\mathbf{C}$. By commutativity of the above diagram, it follows $UH\mu$ is a colimiting cocone of UHD in $\mathbf{C}$. But U reflects colimits, so this means that $H\mu$ is a colimiting cocone of HD, as required.   □

Therefore, in the situation of the above theorem, both functors H and T have parameterised initial algebras by Theorem 2. This brings us to our next theorem.

Theorem 18. *Let H and T be ω-cocontinuous functors, such that the diagram*

$$
\begin{array}{ccc}
\mathbf{C}_a^{n+1} & \xrightarrow{\ U^{\times(n+1)}\ } & \mathbf{C}^{n+1} \\
{\scriptstyle H}\big\downarrow & & \big\downarrow{\scriptstyle T} \\
\mathbf{C}_a & \xrightarrow[\ \ U\ \ ]{} & \mathbf{C}
\end{array}
$$

commutes. Let $(T^\dagger, \phi)$ and $(H^\dagger, \psi)$ be their parameterised initial algebras. Then:

1. *The following diagram:*

$$
\begin{array}{ccc}
\mathbf{C}_a^{n} & \xrightarrow{\ U^{\times n}\ } & \mathbf{C}^{n} \\
{\scriptstyle H^\dagger}\big\downarrow & & \big\downarrow{\scriptstyle T^\dagger} \\
\mathbf{C}_a & \xrightarrow[\ \ U\ \ ]{} & \mathbf{C}
\end{array}
$$

commutes.

2. *The following (2-categorical) diagram:*

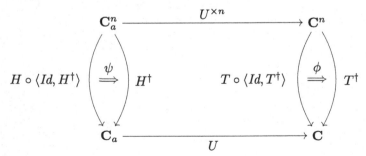

commutes.

Proof. The first statement follows by [13, Corollary 4.21] and the second statement follows by [13, Corollary 4.27].   □

Remark 19. The above theorem shows that the parameterised initial algebras of H and T respect the forgetful functor U and are therefore constructed in the same way.

Remark 20. If one is not interested in interpreting inductive data types defined by mutual induction, then there is no need to form *parameterised* initial algebras, but merely initial algebras. In that case, the assumption that $\mathbf{C}$ has all ω-colimits may be relaxed and one can assume that $\mathbf{C}$ has colimits of the initial sequences of the relevant functors. Then, most of the results presented here can be simplified in a straightforward manner to handle this case.

6   Categorical Model

In this section we formulate our categorical model which we use to interpret **Aff**.

Notation 21. *We write* **DCPO** *(***DCPO**$_{\perp!}$*) for the category of (pointed) dcpo's and (strict) Scott-continuous maps between them.*

Definition 22. *A categorical model of* **Aff** *is given by the following data:*

1. *A symmetric monoidal category* $(\mathbf{C}, \otimes, I, \alpha, \lambda, \rho, \sigma)$.
2. *An initial object* $\varnothing \in \mathrm{Ob}(\mathbf{C})$ *and binary coproducts* $(A + B, \mathit{left}_{A,B}, \mathit{right}_{A,B})$.
3. *The tensor product* $\otimes$ *distributes over* $+$.
4. *For each atomic type* $\mathbf{A} \in \mathcal{A}$, *an object* $\mathbf{A} \in \mathrm{Ob}(\mathbf{C})$ *together with a discarding map* $\diamond_{\mathbf{A}} : \mathbf{A} \to I$.
5. *The category* $\mathbf{C}$ *has all* ω-*colimits and* $\otimes$ *is an* ω-*cocontinuous functor.*
6. *The category* $\mathbf{C}$ *is* **DCPO**$_{\perp!}$-*enriched with least morphisms denoted* $\perp_{A,B}$ *and such that the symmetric monoidal structure and the coproduct structure are both* **DCPO**-*enriched.*

This data suffices to interpret the language in the following way:

1. To interpret pair types.
2. To interpret sum types.
3. Used in the interpretation of **while** loops.
4. Necessary for the affine interpretation of the language.
5. To interpret inductive data types by forming parameterised initial algebras.
6. Used in the interpretation of **while** loops.

Assumption 23. *Henceforth,* $\mathbf{C}$ *refers to an arbitrary, but fixed, model of* **Aff** *and* $\mathbf{C}_a := \mathbf{C}/I$ *refers to the corresponding slice category with the tensor unit* I.

By using results from Sect. 5, we can now easily establish some important properties of the category $\mathbf{C}_a$. By Proposition 13, it follows $\mathbf{C}_a$ has a symmetric monoidal structure with tensor product $\otimes_a$ and by Proposition 14, it follows $\mathbf{C}_a$ has finite coproducts with coproduct functor $+_a$. We also know $\mathbf{C}_a$ has ω-colimits by Proposition 16. Finally, the next proposition is crucial for the construction of discarding maps for inductive data types.

Proposition 24. *The functors* $\otimes_a : \mathbf{C}_a \times \mathbf{C}_a \to \mathbf{C}_a$ *and* $+_a : \mathbf{C}_a \times \mathbf{C}_a \to \mathbf{C}_a$ *are both ω-cocontinuous.*

Proof. In the previous section we showed $\odot \circ (U \times U) = U \circ \odot_a : \mathbf{C}_a \times \mathbf{C}_a \to \mathbf{C}$, for $\odot \in \{\otimes, +\}$. Then, by Theorem 17, it follows $\odot_a$ is also ω-cocontinuous. $\square$

Therefore, by Theorem 2, we see that both categories $\mathbf{C}$ and $\mathbf{C}_a$ have sufficient structure to form parameterised initial algebras for all functors composed out of tensors, coproducts and constants. The category $\mathbf{C}_a$ has the additional benefit that its parameterised initial algebras also come equipped with discarding maps.

6.1   Concrete Models

In this subsection we consider some concrete models of **Aff**.

Example 25. The terminal category **1** is a (completely degenerate) **Aff** model.

Next, we consider some non-degenerate models.

Example 26. The category $\mathbf{DCPO}_{\perp!}$ is an **Aff** model.

However, in this model every object has a canonical comonoid structure, so it is not a truly representative model for an affine type system like ours. In the next example we describe a more representative model which has been studied in the context of circuit description languages and quantum programming.

Example 27. Let $\mathbf{M}$ be a small $\mathbf{DCPO}_{\perp!}$-symmetric monoidal category and let $\widehat{\mathbf{M}} = [\mathbf{M}^{\mathrm{op}}, \mathbf{DCPO}_{\perp!}]$ be the indicated $\mathbf{DCPO}_{\perp!}$-functor category. Then $\widehat{\mathbf{M}}$ is an **Aff** model when equipped with the Day convolution monoidal structure [3]. By making suitable choices for $\mathbf{M}$, the category $\widehat{\mathbf{M}}$ becomes a model of Proto-Quipper-M [18] and ECLNL [11,12], which are programming languages for string diagrams that have also been studied in the context of quantum programming.

Next, we discuss how fragments of the language may be interpreted in categories of W*-algebras [22], which are used to study quantum computing.

Example 28. Let $\mathbf{W}^*_{\mathrm{NMIU}}$ be the category of W*-algebras and normal unital *-homomorphisms between them. Let $\mathbf{V} := (\mathbf{W}^*_{\mathrm{NMIU}})^{\mathrm{op}}$ be its opposite category. Then $\mathbf{V}$ is an **Aff** model without recursion [7], i.e., one can interpret all **Aff** constructs except for while loops within $\mathbf{V}$, because $\mathbf{V}$ is not $\mathbf{DCPO}_{\perp!}$-enriched.

Example 29. Let $\mathbf{W}^*_{\mathrm{NCPSU}}$ be the category of W*-algebras and normal completely-positive subunital maps between them. Let $\mathbf{D} := (\mathbf{W}^*_{\mathrm{NCPSU}})^{\mathrm{op}}$ be its opposite category. Then $\mathbf{D}$ is an **Aff** model which supports simple non-nested type induction, i.e., one can interpret all **Aff** constructs within $\mathbf{D}$ using the methods described in this paper, provided that inductive data types contain at most one free type variable.

Remark 30. In fact, it is possible to interpret all of QPL (and therefore also **Aff** which is a fragment of QPL) by using an adjunction between $\mathbf{V}$ and $\mathbf{D}$, as was shown in [17]. However, this requires considering the specifics of this particular model, which has not been axiomatised yet, and separating the interpretation of types and values (in $\mathbf{V}$) from the interpretation of terms (in $\mathbf{D}$).

$$\begin{array}{ll}
[\![\Theta \vdash A]\!] : \mathbf{C}^{|\Theta|} \to \mathbf{C} & [\![\Theta \vdash A]\!] : \mathbf{C}_a^{|\Theta|} \to \mathbf{C}_a \\
[\![\Theta \vdash \Theta_i]\!] = \Pi_i & [\![\Theta \vdash \Theta_i]\!] = \Pi_i \\
[\![\Theta \vdash I]\!] = K_I & [\![\Theta \vdash I]\!] = K_{(I, \mathrm{id}_I)} \\
[\![\Theta \vdash \mathbf{A}]\!] = K_{\mathbf{A}} & [\![\Theta \vdash \mathbf{A}]\!] = K_{(\mathbf{A}, \diamond_{\mathbf{A}})} \\
[\![\Theta \vdash A + B]\!] = + \circ \langle [\![\Theta \vdash A]\!], [\![\Theta \vdash B]\!] \rangle & [\![\Theta \vdash A + B]\!] = +_a \circ \langle [\![\Theta \vdash A]\!], [\![\Theta \vdash B]\!] \rangle \\
[\![\Theta \vdash A \otimes B]\!] = \otimes \circ \langle [\![\Theta \vdash A]\!], [\![\Theta \vdash B]\!] \rangle & [\![\Theta \vdash A \otimes B]\!] = \otimes_a \circ \langle [\![\Theta \vdash A]\!], [\![\Theta \vdash B]\!] \rangle \\
[\![\Theta \vdash \mu X.A]\!] = [\![\Theta, X \vdash A]\!]^{\dagger} & [\![\Theta \vdash \mu X.A]\!] = [\![\Theta, X \vdash A]\!]^{\dagger}
\end{array}$$

Fig. 4. Standard (left) and affine (right) interpretations of types.

7 Denotational Semantics of Aff

In this section we present the denotational semantics of **Aff**. First, we show how types are interpreted in Sect. 7.1. Since our type system is affine, we construct discarding maps for all types in Sect. 7.2. Folding and unfolding of inductive types are shown to be discardable operations in Sect. 7.3. The interpretations of terms and configurations are defined in Sect. 7.4 and Sect. 7.5. Finally, we prove soundness and adequacy in Sect. 7.6.

7.1 Interpretation of Types

The (standard) interpretation of a type $\Theta \vdash A$ is a functor $[\![\Theta \vdash A]\!] : \mathbf{C}^{|\Theta|} \to \mathbf{C}$, defined in Fig. 4 (left), where K_X indicates the constant X-functor. We begin by showing that this assignment is well-defined, i.e., we have to show that the required parameterised initial algebras exist.

Proposition 31. $[\![\Theta \vdash A]\!]$ *is a well-defined ω-cocontinuous functor.*

Proof. Projection functors and constant functors are obviously ω-cocontinuous. The coproduct functor is ω-cocontinuous, because it is given by a colimiting construction. The tensor product $\otimes$ is ω-cocontinuous by assumption. Also, ω-cocontinuous functors are closed under composition and pairing [9]. By Theorem 2, $[\![\Theta, X \vdash A]\!]^{\dagger}$ is well-defined and also an ω-cocontinuous functor. $\square$

The semantics of terms is defined on closed types, so for brevity we introduce the following notation.

Notation 32. *For any closed type $\cdot \vdash A$, let $[\![A]\!] := [\![\cdot \vdash A]\!](*) \in \mathrm{Ob}(\mathbf{C})$, where $*$ indicates the only object of the terminal category $\mathbf{1}$.*

7.2 Affine Structure of Types

In this subsection we describe the affine structure of our types by constructing an appropriate discarding map for every type. This is achieved by using the results we established in Sect. 5 and by providing an *affine interpretation of types* as functors on the slice category $\mathbf{C}_a = \mathbf{C}/I$. The affine interpretation is related to the standard one via the forgetful functor which results in the construction of the required discarding maps.

The affine interpretation of a type $\Theta \vdash A$ is a functor $[\![\Theta \vdash A]\!] : \mathbf{C}_a^{|\Theta|} \to \mathbf{C}_a$, defined in Fig. 4 (right).

Proposition 33. *$[\![\Theta \vdash A]\!]$ is a well-defined ω-cocontinuous functor.*

Proof. The tensor product $\otimes_a$ and coproduct functors $+_a$ are ω-cocontinuous by Proposition 24. Using the same arguments as in Proposition 31, we finish the proof.   □

Notation 34. *For any closed type $\cdot \vdash A$, let $[\![A]\!] := [\![\cdot \vdash A]\!](*) \in \mathrm{Ob}(\mathbf{C}_a)$.*

We proceed by describing the relationship between the standard and affine interpretation of types.

Theorem 35. *For any type $\Theta \vdash A$, the following diagram*

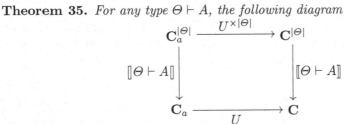

commutes. Therefore, for any closed type $\cdot \vdash A$, we have $[\![A]\!] = U[\![A]\!]$.

Proof. By induction on $\Theta \vdash A$ using the established results from Sect. 5.   □

This theorem shows that for any closed type A, we have $[\![A]\!] = ([\![A]\!], \diamond_{[\![A]\!]})$, where the discarding map $\diamond_{[\![A]\!]} : [\![A]\!] \to I$ is constructed by the affine type interpretation in Fig. 4 (right). We will later see (Theorem 39) that the interpretations of our values are discardable morphisms with respect to this choice of discarding maps.

7.3   Folding and Unfolding of Inductive Data Types

The purpose of this subsection is to define *folding* and *unfolding* of inductive data types (which we need to define the term semantics) and also to demonstrate that folding/unfolding is a discardable isomorphism with respect to the affine structure of our types.

Lemma 36. (Type Substitution). *Let $\Theta, X \vdash A$ and $\Theta \vdash B$ be types. Then:*

1. $[\![\Theta \vdash A[B/X]]\!] = [\![\Theta, X \vdash A]\!] \circ \langle Id, [\![\Theta \vdash B]\!]\rangle.$
2. $[\![\Theta \vdash A[B/X]]\!] = [\![\Theta, X \vdash A]\!] \circ \langle Id, [\![\Theta \vdash B]\!]\rangle.$

Proof. Straightforward induction, essentially the same as [13, Lemma 6.5].   □

Definition 37. *For any closed type* $\cdot \vdash \mu X.A$, *we define two isomorphisms:*

$$\text{fold}_{\mu X.A} : [\![A[\mu X.A/X]]\!] = [\![X \vdash A]\!][\![\mu X.A]\!] \cong [\![\mu X.A]\!] : \text{unfold}_{\mu X.A}$$
$$\text{fold}_{\mu X.A} : [\![A[\mu X.A/X]]\!] = [\![X \vdash A]\!][\![\mu X.A]\!] \cong [\![\mu X.A]\!] : \text{unfold}_{\mu X.A}$$

Since type substitution holds up to equality, it follows that folding/unfolding of inductive data types is determined entirely by the initial algebra structure of the corresponding endofunctors. Finally, we show that folding/unfolding of an inductive data type is the same isomorphism for both the standard and affine type interpretations.

Theorem 38. *Given a closed type* $\cdot \vdash \mu X.A$, *then the following diagram*

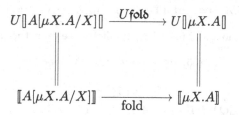

commutes.

Proof. This follows immediately by Theorem 18 (2).   □

Therefore folding/unfolding of types is a discardable isomorphism.

7.4   Interpretation of Terms

In this subsection we explain how to interpret the terms of **Aff**.

A variable context $\Gamma = x_1 : A_1, \ldots, x_n : A_n$ is interpreted as the object $[\![\Gamma]\!] := [\![A_1]\!] \otimes \cdots \otimes [\![A_n]\!] \in \text{Ob}(\mathbf{C})$. A term judgement $\vdash \langle \Gamma \rangle\ M\ \langle \Sigma \rangle$ is interpreted as a morphism $[\![\vdash \langle \Gamma \rangle\ M\ \langle \Sigma \rangle]\!] : [\![\Gamma]\!] \to [\![\Sigma]\!]$ of $\mathbf{C}$ which is defined in Fig. 5. For brevity, we will simply write $[\![M]\!] := [\![\vdash \langle \Gamma \rangle\ M\ \langle \Sigma \rangle]\!]$ whenever the contexts are clear. Next, we clarify some of the notation used in Fig. 5. The map $\diamond_{[\![A]\!]}$ is defined in Sect. 7.2, as already explained. In order to interpret **while** loops, we use a Scott-continuous endofunction W_f, which is defined as follows. For a morphism $f : A \otimes \mathbf{bit} \to A \otimes \mathbf{bit}$, we set:

$$W_f : \mathbf{C}\,(A \otimes \mathbf{bit}, A \otimes \mathbf{bit}) \to \mathbf{C}(A \otimes \mathbf{bit}, A \otimes \mathbf{bit})$$
$$W_f(g) = \big[\text{id} \otimes \text{left}_{I,I},\ g \circ f \circ (\text{id} \otimes \text{right}_{I,I})\big] \circ d_{A,I,I},$$

where $d_{A,I,I} : A \otimes (I + I) \to (A \otimes I) + (A \otimes I)$ is the isomorphism which is induced by the distributivity of $\otimes$ over $+$ (see Definition 22). Finally, for a pointed dcpo D and Scott-continuous endofunction $h : D \to D$, the *least fixpoint* of h is given by $\text{lfp}(h) := \bigvee_{i=0}^{\infty} h^i(\bot)$, where $\bot$ is the least element of D.

$$[\![\vdash \langle \Gamma \rangle \text{ new unit } u \ \langle \Gamma, u : I \rangle]\!] := r^{-1}$$

$$[\![\vdash \langle \Gamma, x : A \rangle \text{ discard } x \ \langle \Gamma \rangle]\!] := r \circ (\text{id} \otimes \diamond)$$

$$[\![\vdash \langle \Gamma \rangle \ M; N \ \langle \Sigma \rangle]\!] := [\![N]\!] \circ [\![M]\!]$$

$$[\![\vdash \langle \Gamma \rangle \text{ skip } \langle \Gamma \rangle]\!] := \text{id}$$

$$[\![\vdash \langle \Gamma, b : \text{bit} \rangle \text{ while } b \text{ do } M \ \langle \Gamma, b : \text{bit} \rangle]\!] := \text{lfp}(W_{[\![M]\!]})$$

$$[\![\vdash \langle \Gamma, x : A \rangle \ y = \text{left}_{A,B} \ x \ \langle \Gamma, y : A + B \rangle]\!] := \text{id} \otimes \text{left}_{A,B}$$

$$[\![\vdash \langle \Gamma, x : B \rangle \ y = \text{right}_{A,B} \ x \ \langle \Gamma, y : A + B \rangle]\!] := \text{id} \otimes \text{right}_{A,B}$$

$$[\![\vdash \langle \Gamma, y : A + B \rangle \text{ case } y \text{ of } \{\text{left } x_1 \rightarrow M_1 \ |$$
$$\text{right } x_2 \rightarrow M_2\} \ \langle \Sigma \rangle]\!] := [[\![M_1]\!], [\![M_2]\!]] \circ d$$

$$[\![\vdash \langle \Gamma, x_1 : A, x_2 : B \rangle \ x = (x_1, x_2) \ \langle \Gamma, x : A \otimes B \rangle]\!] := \text{id}$$

$$[\![\vdash \langle \Gamma, x : A \otimes B \rangle \ (x_1, x_2) = x \ \langle \Gamma, x_1 : A, x_2 : B \rangle]\!] := \text{id}$$

$$[\![\vdash \langle \Gamma, x : A[\mu X.A/X] \rangle \ y = \text{fold } x \ \langle \Gamma, y : \mu X.A \rangle]\!] := \text{id} \otimes \text{fold}$$

$$[\![\vdash \langle \Gamma, x : \mu X.A \rangle \ y = \text{unfold } x \ \langle \Gamma, y : A[\mu X.A/X] \rangle]\!] := \text{id} \otimes \text{unfold}$$

where r is the right monoidal unit. For simplicity, we omit the monoidal associator.

Fig. 5. Interpretation of **Aff** terms.

$$[\![\cdot \vdash * : I]\!] := \text{id}_I$$

$$[\![Q \vdash \text{left}_{A,B}v : A + B]\!] := \text{left} \circ [\![v]\!]$$

$$[\![Q \vdash \text{right}_{A,B}v : A + B]\!] := \text{right} \circ [\![v]\!]$$

$$[\![Q_1, Q_2 \vdash (v, w) : A \otimes B]\!] := ([\![v]\!] \otimes [\![w]\!]) \circ \lambda_I^{-1}$$

$$[\![Q \vdash \text{fold}_{\mu X.A}v : \mu X.A]\!] := \text{fold} \circ [\![v]\!]$$

Fig. 6. Interpretation of **Aff** values.

7.5 Interpretation of Configurations

Before we explain how to interpret configurations, we have to show how to interpret values.

Interpretation of Values. The interpretation of a value $\vdash v : A$ is a morphism $[\![\vdash v : A]\!] : I \rightarrow [\![A]\!]$, and we shall simply write $[\![v]\!]$ if its type is clear from context. The interpretation is defined in Fig. 6. In order to prove soundness of our *affine* type system, we have to show every value is discardable.

Theorem 39. *For every value $\vdash v : A$, we have:* $\diamond_{[\![A]\!]} \circ [\![\vdash v : A]\!] = id_I$.

Proof. By construction, $\diamond_{[\![A]\!]}$ enjoys all of the properties established in Sect. 5. The proof proceeds by induction on the derivation of $\vdash v : A$. The base case is trivial. Discardable morphisms are closed under composition, because $\mathbf{C}_a$ is a category. Moreover, discardable maps are closed under tensor products (Proposition 13). Using the induction hypothesis, it suffices to show that the coproduct injections and folding are discardable maps. But this follows by Proposition 14 and Theorem 38, respectively.   □

Given a variable context $\Gamma = x_1 : A_1, \ldots, x_n : A_n$, then a value context $\Gamma \vdash V$ is interpreted by the morphism:

$$[\![\Gamma \vdash V]\!] = \left(I \xrightarrow{\cong} I^{\otimes n} \xrightarrow{[\![v_1]\!] \otimes \cdots \otimes [\![v_n]\!]} [\![\Gamma]\!] \right),$$

where $V = \{x_1 = v_1, \ldots, x_n = v_n\}$ and we abbreviate this by writing $[\![V]\!]$. Note that $[\![V]\!]$ is also discardable due to Theorem 39.

Interpretation of Configurations. A configuration $\Gamma; \Sigma \vdash (M \mid V)$ is interpreted as the morphism

$$[\![\Gamma; \Sigma \vdash (M \mid V)]\!] = \left(I \xrightarrow{[\![\Gamma \vdash V]\!]} [\![\Gamma]\!] \xrightarrow{[\![\vdash \langle \Gamma \rangle \; M \; \langle \Sigma \rangle]\!]} [\![\Sigma]\!] \right).$$

We write $[\![(M \mid V)]\!]$ for this morphism whenever the contexts are clear.

7.6 Soundness and Computational Adequacy

Soundness is the statement that the denotational semantics is invariant under program execution.

Theorem 40. (Soundness). *If $C \rightsquigarrow D$, then $[\![C]\!] = [\![D]\!]$.*

Proof. Straightforward induction. □

We conclude our technical contributions by proving a computational adequacy result. Towards this end, we have to assume that our categorical model is not degenerate.

Definition 41. *A computationally adequate **Aff** model is an **Aff** model, where $id_I \neq \bot$.*

A closed term $\vdash \langle \cdot \rangle \; M \; \langle \Sigma \rangle$ is said to *terminate*, denoted $M \Downarrow$, if there exists a terminal configuration T, such that $(M \mid \cdot) \rightsquigarrow_* T$, where $\rightsquigarrow_*$ is the reflexive and transitive closure of $\rightsquigarrow$.

Theorem 42. (Adequacy). *Let $\vdash \langle \cdot \rangle \; M \; \langle \Sigma \rangle$ be a closed term. Then:*

$$[\![M]\!] \neq \bot \;\; iff \; M \Downarrow .$$

Proof. In Appendix A. □

8 Future Work

As part of future work it will be interesting to see whether these methods can be adapted to also work with coinductive data types and/or with recursive data types where function types ($\multimap$) become admissible constructs within the type recursion schemes. This is certainly a more challenging problem which would probably require us to assume additional structure within the model, such as a limit-colimit coincidence [21], so that we may deal with the mixed-variance induced by $\multimap$ in the interpretation of our types. Aside from this, the mixed-variance induced by function types poses further problems for the slice construction and it is likely that we would have to modify it, so that it can simultaneously accommodate the addition of limits while also handling mixed-variance functors as well.

9   Conclusion and Related Work

Since the introduction of Linear Logic [5], there has been a massive amount of research into finding suitable models for (fragments) of Linear Logic (see [14] for an excellent overview). However, there has been less research into models of affine logics and affine type systems. The principle difference between linear and affine logic is that weakening is restricted in the former, but allowed in the latter, so affine models have to contain additional discarding maps.

Most models of affine type systems (that I am aware of) use some specific properties of the model, such as finding a suitable (sub)category where the tensor unit I is a terminal object, in order to construct the required discarding maps (e.g. [2,6,8,17,20]). This means, the solution is provided directly by the *model*. In this paper, we have taken a different approach, because we present a *semantic* solution to this problem. The only assumption that we have made for our model is that the interpretations of the atomic types are equipped with suitable discarding maps and we then show how to construct all other required discarding maps by considering an additional and non-standard interpretation of types within a slice category.

Overall, the "model" solution is certainly simpler and more concise compared to the "semantic" solution presented here. On the other hand, our solution in this paper is very general and can in principle be applied to models where the required discarding maps are unknown a priori.

Acknowledgements. I thank Romain Péchoux, Simon Perdrix and Mathys Rennela for discussions about the methods in this paper. I also thank the anonymous referees for the very useful feedback they provided. Finally, I gratefully acknowledge financial support from the French projects ANR-17-CE25-0009 SoftQPro and PIA-GDN/Quantex.

A   Proof of Theorem 42

As a simple consequence of Soundness (Theorem 40), we get the following corollary.

Corollary 43. *Let* $\vdash \langle \cdot \rangle\ M\ \langle \Sigma \rangle$ *be a closed term. If* $M \Downarrow$ *then* $\llbracket M \rrbracket \neq \bot$ *.*

Proof. Assume for contradiction $\llbracket M \rrbracket = \bot$. Let $M' \equiv M;$ **discard** Σ be the program that discards all output variables of M. Clearly, $(M' \mid \cdot) \leadsto_* (\mathbf{skip} \mid \cdot)$ and therefore $\mathrm{id}_I = \llbracket M' \rrbracket = \diamond \circ \llbracket M \rrbracket = \bot$, which contradicts Definition 41.   □

In the remainder of this appendix, we will show that the converse implication also holds which therefore completes the proof of Theorem 42. Our proof strategy is a simplification of the proof strategy of [17] which is in turn based on the proof strategy of [15].

We provide a brief outline of our proof strategy. In Subsect. A.1, we extend the language with finitary (or bounded) primitives for recursion and show that our extension preserves all of the fundamental properties of the language stated so far (Theorem 44). Any finitary configuration terminates in the extended language (Lemma 45) which allows us to easily prove a finitary version of adequacy (Corrolary 46). We then show that finitary configurations approximate ordinary ones both operationally (Lemma 48) and denotationally (Lemma 49) which then allows us to finish the proof.

A.1   Language Extension

We extend the term language by adding two new terms:

$$\frac{}{\vdash \langle \Gamma \rangle \; \bot_{\Gamma,\Sigma} \; \langle \Sigma \rangle} \qquad \frac{\vdash \langle \Gamma, b : \mathbf{bit} \rangle \; M \; \langle \Gamma, b : \mathbf{bit} \rangle}{\vdash \langle \Gamma, b : \mathbf{bit} \rangle \; \mathbf{while}^n \; b \; \mathbf{do} \; M \; \langle \Gamma, b : \mathbf{bit} \rangle}$$

where $n \geq 0$ ranges over the natural numbers. Well-formed configuration are defined in the same way as before. We extend the operational semantics by adding the following rules:

$$\frac{}{(\bot; P \mid V) \rightsquigarrow (\bot \mid V)} \qquad \frac{}{(\mathbf{while}^0 \; b \; \mathbf{do} \; M \mid V, b = v) \rightsquigarrow (\bot \mid V, b = v)}$$

$$\frac{}{(\mathbf{while}^{n+1} \; b \; \mathbf{do} \; M \mid V, b = v) \rightsquigarrow (\mathbf{if} \; b \; \mathbf{then} \; \{M; \mathbf{while}^n \; b \; \mathbf{do} \; M\} \; \mathbf{else} \; \mathbf{skip} \mid V, b = v)}$$

We extend the denotational semantics by adding interpretations for the newly added terms:

$$[\![\vdash \langle \Gamma \rangle \; \bot_{\Gamma,\Sigma} \; \langle \Sigma \rangle]\!] := \bot_{[\![\Gamma]\!],[\![\Sigma]\!]}$$
$$[\![\vdash \langle \Gamma, b : \mathbf{bit} \rangle \; \mathbf{while}^n \; b \; \mathbf{do} \; M \; \langle \Gamma, b : \mathbf{bit} \rangle]\!] := W^n_{[\![M]\!]}(\bot)$$

Configurations are interpreted as before. We update the notion of terminal configuration to also include configurations of the form $(\bot \mid V)$.

Theorem 44. ([17]). *Subject Reduction (Theorem 6), Progress (Theorem 8) and Soundness (Theorem 40) also hold true for the extended language (using the updated language notions).*

A.2   The Proof

We shall say that a term is *finitary* if it does not contain any unindexed while loops and we shall say a term is *ordinary* if it does not contain any indexed while loops or $\perp_{\Gamma,\Sigma}$ subterms. So, an ordinary term is simply a term in the ordinary (unextended) language. Similarly, a *finitary (ordinary) configuration* is a configuration $(M \mid V)$ where M is finitary (ordinary). A finitary configuration is called as such, because all of its reduction sequences terminate.

Lemma 45. ([17]). *For any finitary configuration $\mathcal{C}$, there exists $n \in \mathbb{N}$, such that the length of every reduction sequence from $\mathcal{C}$ is at most n.*

Note that, every terminal finitary configuration $\mathcal{T}$ has to be of the form $(\mathbf{skip} \mid V)$, in which case we say it is a $\mathbf{skip}$-*configuration*, or of the form $(\perp \mid V)$, in which case we say $\mathcal{T}$ is a $\perp$-*configuration*. For any closed term $\vdash \langle \cdot \rangle\, M\, \langle \Sigma \rangle$ in the extended language, we shall write $M \Downarrow_o$, if M terminates in the ordinary sense, i.e., if $(M \mid \cdot) \rightsquigarrow_* (\mathbf{skip} \mid V)$, for some value assignment V.

Corollary 46. *For any closed finitary term M, if $[\![M]\!] \neq \perp$ then $M \Downarrow_o$.*

Proof. By Lemma 45, we know $(M \mid \cdot) \rightsquigarrow_* (T' \mid V)$, where T' is either $\mathbf{skip}$ or $\perp$. If $T' = \perp$, then by soundness, it follows $[\![M]\!] = \perp$. Therefore $M \Downarrow_o$. $\qquad\square$

We proceed by defining an approximation relation $(-\, \blacktriangleleft\, -)$ between finitary terms and ordinary terms. It is the smallest relation that satisfies the rules:

$$\frac{M' \blacktriangleleft M}{\mathbf{while}^n\, b\, \mathbf{do}\, M' \blacktriangleleft \mathbf{while}\, b\, \mathbf{do}\, M} \qquad \frac{M' \blacktriangleleft M \qquad N' \blacktriangleleft N}{M'; N' \blacktriangleleft M; N}$$

$$\frac{M_1' \blacktriangleleft M_1 \qquad M_2' \blacktriangleleft M_2}{\mathbf{case}\, y\, \mathbf{of}\, \{\mathbf{left}_{A,B}\, x_1 \to M_1' \mid \mathbf{right}_{A,B}\, x_2 \to M_2' \}\, \blacktriangleleft} \\ \mathbf{case}\, y\, \mathbf{of}\, \{\mathbf{left}_{A,B}\, x_1 \to M_1 \mid \mathbf{right}_{A,B}\, x_2 \to M_2 \}$$

$$\frac{}{M \blacktriangleleft M} \text{ all other terms except } \perp.$$

We also define an approximation relation $(-\, \sqsubset\, -)$ between finitary configurations and ordinary configurations to be the smallest relation satisfying the rule:

$$\frac{M' \blacktriangleleft M}{(M' \mid V) \sqsubset (M \mid V)}$$

Remark 47. If $(M' \mid V) \sqsubset (M \mid V)$, then M and M' do not contain any $\perp$ subterms.

Next, we show the relation $(-\, \sqsubset\, -)$ approximates the ordinary operational semantics.

Lemma 48. *Let $\mathcal{C}$ be an ordinary configuration and $\mathcal{C}'$ a finitary configuration with $\mathcal{C}' \sqsubset \mathcal{C}$. Assume that $\mathcal{C}' \rightsquigarrow_* \mathcal{T}$, where $\mathcal{T}$ is a $\mathbf{skip}$-configuration. Then $\mathcal{C} \rightsquigarrow_* \mathcal{T}$.*

Proof. This is just a simple special case of [17, Lemma 40]. $\qquad\square$

The above lemma shows that if a finitary configuration terminates in the ordinary sense, then so does the ordinary configuration which is approximated by it. Next we show that ordinary configurations are also approximated by finitary configurations in a denotational sense.

Lemma 49. ([17]). *For any ordinary term M and configuration C:*

$$[\![M]\!] = \bigvee_{M' \blacktriangleleft M} [\![M']\!] \qquad\qquad [\![C]\!] = \bigvee_{C' \sqsubseteq C} [\![C']\!].$$

Finally, we can state the proof of Theorem 42.

Proof of Theorem 42. By Corollary 43, it suffices to show that if $[\![M]\!] \neq \bot$ then $M \Downarrow_o$. By Lemma 49, we have $[\![M]\!] = \bigvee_{M' \blacktriangleleft M} [\![M']\!]$. Since $[\![M]\!] \neq \bot$, then it follows there exists a finitary term $M' \blacktriangleleft M$, such that $[\![M']\!] \neq \bot$. By Corollary 46, it follows $M' \Downarrow_o$ and by Lemma 48 it follows $M \Downarrow_o$. $\square$

References

1. Borceux, F.: Handbook of Categorical Algebra: Volume 1, Basic Category Theory, vol. 1. Cambridge University Press, Cambridge (1994)
2. Clairambault, P., de Visme, M., Winskel, G.: Game semantics for quantum programming. PACMPL **3**(POPL), 321–3229 (2019). https://doi.org/10.1145/3290345
3. Day, B.: On closed categories of functors II. In: Kelly, G.M. (ed.) Category Seminar. LNM, vol. 420, pp. 20–54. Springer, Heidelberg (1974). https://doi.org/10.1007/BFb0063098
4. Fiore, M.P.: Axiomatic domain theory in categories of partial maps. Ph.D. thesis, University of Edinburgh, UK (1994)
5. Girard, J.: Linear logic. Theor. Comput. Sci. **50**, 1–102 (1987). https://doi.org/10.1016/0304-3975(87)90045-4
6. Jacobs, B.: Semantics of lambda-I and of other substructure lambda calculi. In: Bezem, M., Groote, J.F. (eds.) TLCA 1993. LNCS, vol. 664, pp. 195–208. Springer, Heidelberg (1993). https://doi.org/10.1007/BFb0037107
7. Kornell, A.: Quantum collections. Int. J. Math. **28**(12), 1750085 (2017). https://doi.org/10.1142/S0129167X17500859
8. Laird, J.: A game semantics of linearly used continuations. In: Gordon, A.D. (ed.) FoSSaCS 2003. LNCS, vol. 2620, pp. 313–327. Springer, Heidelberg (2003). https://doi.org/10.1007/3-540-36576-1_20
9. Lehmann, D.J., Smyth, M.B.: Algebraic specification of data types: a synthetic approach. Math. Syst. Theory **14**, 97–135 (1981). https://doi.org/10.1007/BF01752392
10. Lindenhovius, B., Mislove, M., Zamdzhiev, V.: Mixed linear and non-linear recursive types. Proc. ACM Program. Lang. **3**(ICFP), 1111–11129 (2019). https://doi.org/10.1145/3341715
11. Lindenhovius, B., Mislove, M., Zamdzhiev, V.: Semantics for a lambda calculus for string diagrams (2020). preprint

12. Lindenhovius, B., Mislove, M.W., Zamdzhiev, V.: Enriching a Linear/non-linear Lambda Calculus: a programming language for string diagrams. In: Dawar, A., Grädel, E. (eds.) Proceedings of the 33rd Annual ACM/IEEE Symposium on Logic in Computer Science, LICS 2018, Oxford, UK, 09–12 July 2018, pp. 659–668. ACM (2018). https://doi.org/10.1145/3209108.3209196
13. Lindenhovius, B., Mislove, M.W., Zamdzhiev, V.: LNL-FPC: the linear/non-linear fixpoint calculus (2020). preprint. http://arxiv.org/abs/1906.09503
14. Mellies, P.A.: Categorical semantics of linear logic. Panoramas et syntheses **27**, 15–215 (2009)
15. Pagani, M., Selinger, P., Valiron, B.: Applying quantitative semantics to higher-order quantum computing. In: Jagannathan, S., Sewell, P. (eds.) The 41st Annual ACM SIGPLAN-SIGACT Symposium on Principles of Programming Languages, POPL 2014, San Diego, CA, USA, 20–21 January 2014. pp. 647–658. ACM (2014). https://doi.org/10.1145/2535838.2535879
16. Péchoux, R., Perdrix, S., Rennela, M., Zamdzhiev, V.: Quantum programming with inductive datatypes (2020). preprint
17. Péchoux, R., Perdrix, S., Rennela, M., Zamdzhiev, V.: Quantum programming with inductive datatypes: causality and affine type theory. FoSSaCS 2020. LNCS, vol. 12077, pp. 562–581. Springer, Cham (2020). https://doi.org/10.1007/978-3-030-45231-5_29
18. Rios, F., Selinger, P.: A categorical model for a quantum circuit description language. In: QPL 2017 (2017). https://doi.org/10.4204/EPTCS.266.11
19. Selinger, P.: Towards a quantum programming language. Math. Struct. Comput. Sci. **14**(4), 527–586 (2004). https://doi.org/10.1017/S0960129504004256
20. Selinger, P., Valiron, B.: Quantum Lambda Calculus (2009). https://doi.org/10.1017/CBO9781139193313.005
21. Smyth, M.B., Plotkin, G.D.: The category-theoretic solution of recursive domain equations. SIAM J. Comput. **11**(4), 761–783 (1982). https://doi.org/10.1137/0211062
22. Takesaki, M.: Theory of Operator Algebras, vol. I. II and III. Springer, Heidelberg (2002). https://doi.org/10.1007/978-1-4612-6188-9
23. Wootters, W.K., Zurek, W.H.: A single quantum cannot be cloned. Nature **299**(5886), 802–803 (1982)

Author Index

Printed in the United States
By Bookmasters